PROBABILITY AND RANDOM NUMBER

A FIRST GUIDE TO RANDOMNESS

PROBABILITY AND RANDOM NUMBER

A FIRST GUIDE TO RANDOMNESS

HIROSHI SUGITA
Osaka University, Japan

NEW JERSEY • LONDON • SINGAPORE • BEIJING • SHANGHAI • HONG KONG • TAIPEI • CHENNAI • TOKYO

Published by

World Scientific Publishing Co. Pte. Ltd.
5 Toh Tuck Link, Singapore 596224
USA office: 27 Warren Street, Suite 401-402, Hackensack, NJ 07601
UK office: 57 Shelton Street, Covent Garden, London WC2H 9HE

British Library Cataloguing-in-Publication Data
A catalogue record for this book is available from the British Library.

PROBABILITY AND RANDOM NUMBER
A First Guide to Randomness

ISBN 978-981-3228-25-2

Printed in Singapore

Preface

Imagine old times when the word *probability* did not exist. Facing difficult situations that could be expressed as irregular, unpredictable, random, etc. (in what follows, we call it *random*), people were helpless. After a long time, they have found how to describe randomness, how to analyze it, how to define it, and how to make use of it. What is really amazing is that these all have been done in the very rigorous mathematics—just like geometry and algebra.

At high school, students calculate probabilities by counting the number of permutations, combinations, etc. At university, counting is also the most basic method to study probability. The only difference is that we count *huge numbers* at university; e.g., we ask how large $10,000!$ is. To count huge numbers, *calculus*—differentiation and integration—is useful. To deal with extremely huge numbers, taking the limit to infinity often makes things simpler, in which case, calculus is again useful. In short words, at university, counting huge numbers by calculus is the most basic method to study probability.

Why do we count huge numbers? It is because we want to find as many *limit theorems* as possible. Limit theorems are very useful to solve practical problems; e.g., to analyze statistical data of 10,000 people, or to study properties of a certain substance consisting of 6.02×10^{23} molecules. They have a yet more important mission. It is to unlock the secrets of randomness, which is the ultimate aim of studying probability.

What is randomness? Why can limit theorems unlock its secrets? To answer these questions, we feature *random number* as one of the two main subjects of this book.†[1] Without learning random number, we can do

[1] The other one is, of course, *probability*.

calculations and prove theorems about probability, but to understand the essential relation between probability and randomness, the knowledge of random number is necessary.

Another reason why we feature random number is that for proper understanding and implementation of the Monte Carlo method, the knowledge of random number is indispensable. The Monte Carlo method is a numerical method to solve mathematical problems by computer-aided sampling of random variables. Thus, not only in theory but also in practice, learning random number is important.

The prerequisite for this book is first-year university calculus. University mathematics is really difficult. There are three elements of the difficulty.

First, the subtle nuance of concepts described by unfamiliar technical terms. For example, a *random variable* needs a *probability space* as a setup, and should be accompanied by a *distribution*, to make sense. In particular, special attention must be paid to terms—such as *event*—that have mathematical meanings other than usual ones.

Secondly, long proofs and complicated calculations. This book includes many of them. They are unavoidable; for it is not easy to obtain important concepts or theorems. In this book, reasoning by many inequalities, which readers may not be used to, appear here and there. We hope readers to follow the logic patiently.

Thirdly, the fact that *infinity* plays an essential role. Since the latter half of the 19th century, mathematics has developed very much by dealing with infinity directly. However, infinity essentially differs from finity[†2] in many respects, and our usual intuition does not work at all for it. Therefore mathematical concepts about infinity cannot help being so delicate that we must be very careful in dealing with them. In this book, we discuss the distinction between countable set and uncountable set, and the rigorous definition of limit.

This book presents well-known basic theorems with proofs that are not seen in usual probability textbooks; for we want readers to learn that a good solution is not always unique. In general, breakthroughs in science have been made by unusual solutions. We hope readers to know more than one proof for every important theorem.

This is an English translation of my Japanese book *Kakuritsu to ransū* published by *Sugakushobo Co.* To that book, Professors Masato Takei and

[2]finiteness

Tetsuya Hattori gave me suggestions for better future publication. Indeed, they were very useful in preparing the present English version. Mr. Shin Yokoyama at Sugakushobo encouraged me to translate the book. Professor Nicolas Bouleau carefully read the translated English manuscript, and gave me valuable comments, which helped me improve it. Dr. Pan Suqi and Ms. Tan Rok Ting at World Scientific Publishing Co. kindly supported me in producing this English version. I am really grateful to all of them.

Osaka, September 2017 Hiroshi Sugita

Notations and symbols

$A := B$ — A is defined by B ($B =: A$ as well).

$P \Longrightarrow Q$ — P implies Q (logical inclusion).

$\square$ — end of proof.

$\mathbb{N} := \{0, 1, 2, \ldots\}$, the set of all non-negative integers.

$\mathbb{N}_+ := \{1, 2, \ldots\}$, the set of all positive integers.

$\mathbb{R} :=$ the set of all real numbers.

$\prod_{i=1}^{n} a_i := a_1 \times \cdots \times a_n$.

$\max[\min] A :=$ the maximum[minimum] value of $A \subset \mathbb{R}$.

$\max_{t \geqq 0} \left[\min_{t \geqq 0}\right] u(t) :=$ the maximum[minimum] value of $u(t)$ over all $t \geqq 0$.

$\lfloor t \rfloor :=$ the largest integer not exceeding $t \geqq 0$ (rounding down).

$\binom{n}{k} := \dfrac{n!}{(n-k)!k!}$.

$a \approx b$ — a and b are approximately equal to each other.

$a \gg b$ — a is much greater than b ($b \ll a$ as well).

$a_n \sim b_n$ — $a_n/b_n \to 1$, as $n \to \infty$.

$\emptyset :=$ the empty set.

$\mathfrak{P}(\Omega) :=$ the set of all subsets of Ω.

$\mathbf{1}_A(x) := 1\ (x \in A), \quad 0\ (x \notin A)$ (the indicator function of A).

$\#A :=$ the number of elements of A.

$A^c :=$ the complement of A.

$A \times B := \{(x, y) \mid x \in A,\ y \in B\}$ (the direct product of A and B).

Table of Greek letters

A, α	alpha	I, ι	iota	$\mathrm{P}, \rho\ (\varrho)$	rho
B, β	beta	K, κ	kappa	$\Sigma, \sigma\ (\varsigma)$	sigma
Γ, γ	gamma	Λ, λ	lambda	T, τ	tau
Δ, δ	delta	M, μ	mu	Υ, υ	upsilon
$\mathrm{E}, \epsilon\ (\varepsilon)$	epsilon	N, ν	nu	$\Phi, \phi\ (\varphi)$	phi
Z, ζ	zeta	Ξ, ξ	xi	X, χ	chi
H, η	eta	O, o	omicron	Ψ, ψ	psi
$\Theta, \theta\ (\vartheta)$	theta	$\Pi, \pi\ (\varpi)$	pi	Ω, ω	omega

Contents

[3]The subsections with (*) can be skipped.

Chapter 1

Mathematics of coin tossing

Tossing a coin many times, record 1 if it comes up Heads and record 0 if it comes up Tails at each coin toss. Then, we get a long sequence consisting of 0 and 1—let us call such a sequence a $\{0,1\}$-*sequence*—that is random. In this chapter, with such random $\{0,1\}$-sequences as material, we study outlines of

- how to describe randomness (Sec. 1.1),
- how to define randomness (Sec. 1.2),
- how to analyze randomness (Sec. 1.3.1), and
- how to make use of randomness (Sec. 1.3.2, Sec. 1.4).

Readers may think that coin tosses are too simple as a random object, but as a matter of fact, virtually all random objects can be mathematically constructed from them (Sec. 1.5.2). Thus analyzing coin tosses means analyzing all random objects.

In this chapter, we present only basic ideas, and do not prove theorems.

1.1 Mathematical model

For example, the concept 'circle' is obtained by abstracting an essence from various round objects in the world. To deal with circle in mathematics, we consider an equation $(x-a)^2+(y-b)^2=c^2$ as a *mathematical model*. Namely, what we call a circle in mathematics is the set of all solutions of this equation

$$\{(x,y)\,|\,(x-a)^2+(y-b)^2=c^2\}.$$

Similarly, to analyze random objects, since we cannot deal with them directly in mathematics, we consider their mathematical models. For example, when we say 'n coin tosses', it does not mean that we toss a real coin

n times, but it means a mathematical model of it, which is described by mathematical expressions in the same way as 'circle'.

Let us consider a mathematical model of '3 coin tosses'. Let $X_i \in \{0,1\}$ be the outcome (Heads = 1 and Tails = 0) of the i-th coin toss. At high school, students learn the probability that the consecutive outcomes of 3 coin tosses are Heads, Tails, Heads, is

$$P(X_1 = 1,\, X_2 = 0,\, X_3 = 1) = \left(\frac{1}{2}\right)^3 = \frac{1}{8}. \tag{1.1}$$

Here, however, the mathematical definitions of P and X_i are not clear. After making them clear, we can call them a mathematical model of 3 coin tosses.

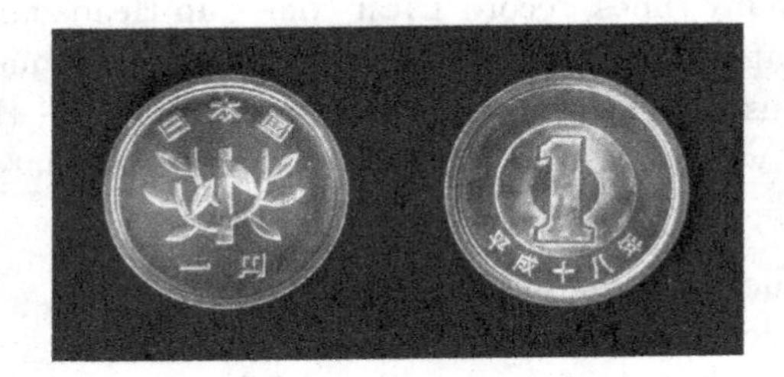

Fig. 1.1 Heads and Tails of 1 JPY coin

Example 1.1. Let $\{0,1\}^3$ denote the set of all $\{0,1\}$-sequences of length 3:

$$\begin{aligned}\{0,1\}^3 &:= \{\,\omega = (\omega_1, \omega_2, \omega_3) \mid \omega_i \in \{0,1\},\ 1 \leqq i \leqq 3\,\} \\ &= \{\,(0,0,0), (0,0,1), (0,1,0), (0,1,1), (1,0,0), \\ &\qquad (1,0,1), (1,1,0), (1,1,1)\,\}.\end{aligned}$$

Let $\mathfrak{P}(\{0,1\}^3)$ be the *power set*[†1] of $\{0,1\}^3$, i.e., the set of all subsets of $\{0,1\}^3$. $A \in \mathfrak{P}(\{0,1\}^3)$ is equivalent to $A \subset \{0,1\}^3$. Let $\#A$ denote the number of elements of A. Now, define a function $P_3 : \mathfrak{P}(\{0,1\}^3) \to [0,1] := \{\,x \mid 0 \leqq x \leqq 1\,\}$ by

$$P_3(A) := \frac{\#A}{\#\{0,1\}^3} = \frac{\#A}{2^3}, \quad A \in \mathfrak{P}(\{0,1\}^3)$$

(See Definition A.2), and functions $\xi_i : \{0,1\}^3 \to \{0,1\},\ \ i = 1,2,3,$ by

$$\xi_i(\omega) := \omega_i, \quad \omega = (\omega_1, \omega_2, \omega_3) \in \{0,1\}^3. \tag{1.2}$$

[1] $\mathfrak{P}$ is the letter P in the *Fractur* typeface.

Each ξ_i is called a *coordinate function*. Then, we have

$$\begin{aligned}
&P_3\left(\left\{\,\omega\in\{0,1\}^3 \;\middle|\; \xi_1(\omega)=1,\ \xi_2(\omega)=0,\ \xi_3(\omega)=1\,\right\}\right)\\
&\qquad = P_3\left(\left\{\,\omega\in\{0,1\}^3 \;\middle|\; \omega_1=1,\ \omega_2=0,\ \omega_3=1\,\right\}\right)\\
&\qquad = P_3(\{\,(1,0,1)\,\}) \;=\; \frac{1}{2^3}. \qquad (1.3)
\end{aligned}$$

Although (1.3) has nothing to do with the real coin tosses, it is formally the same as (1.1). Readers can easily examine the formal identity not only for the case Heads, Tails, Heads, but also for any other possible outcomes of 3 coin tosses. Thus we can compute every probability concerning 3 coin tosses by using P_3 and $\{\xi_i\}_{i=1}^3$. This means that P and $\{X_i\}_{i=1}^3$ in (1.1) can be considered as P_3 and $\{\xi_i\}_{i=1}^3$, respectively. In other words, by the correspondence

$$P \longleftrightarrow P_3, \quad \{X_i\}_{i=1}^3 \longleftrightarrow \{\xi_i\}_{i=1}^3,$$

P_3 and $\{\xi_i\}_{i=1}^3$ are a mathematical model of 3 coin tosses.

The equation $(x-a)^2+(y-b)^2=c^2$ is not a unique mathematical model of 'circle'. There are different models of it; e.g., a parametrized representation

$$\begin{cases} x = c\cos t + a, \\ y = c\sin t + b, \end{cases} \quad 0 \leqq t \leqq 2\pi.$$

You can select suitable mathematical models according with your particular purposes. In the case of coin tosses, it is all the same. We can present another mathematical model of 3 coin tosses.

Example 1.2. (Borel's model of coin tosses) For each $x\in[0,1) := \{\,x \mid 0 \leqq x < 1\}$, let $d_i(x)\in\{0,1\}$ denote the i-th digit of x in its binary expansion (Sec. A.2.2). We write the length of each semi-open interval $[a,b)\subset[0,1)$ as

$$\mathbb{P}(\,[a,b)\,) := b-a.$$

Here, the function $\mathbb{P}$ that returns the lengths of semi-open intervals is called the *Lebesgue measure*. Then, the length of the set of $x\in[0,1)$ for which $d_1(x)$, $d_2(x)$, $d_3(x)$, are 1, 0, 1, respectively, is

$$\begin{aligned}
&\mathbb{P}(\,\{x\in[0,1) \mid d_1(x)=1,\ d_2(x)=0,\ d_3(x)=1\}\,)\\
&\quad = \mathbb{P}\left(\left\{x\in[0,1) \;\middle|\; \frac{1}{2}+\frac{0}{2^2}+\frac{1}{2^3} \leqq x < \frac{1}{2}+\frac{0}{2^2}+\frac{1}{2^3}+\frac{1}{2^3}\right\}\right)\\
&\quad = \mathbb{P}\left(\left[\frac{5}{8},\frac{6}{8}\right)\right) = \frac{1}{8}.
\end{aligned}$$

In the number line with binary scale, it is expressed as a segment:

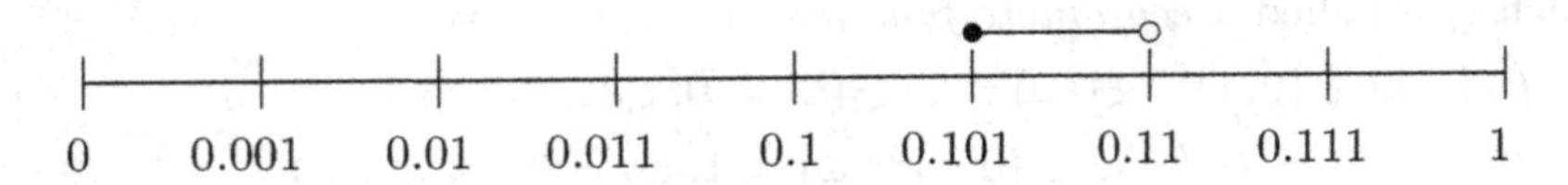

Under the correspondence

$$P \longleftrightarrow \mathbb{P}, \quad \{X_i\}_{i=1}^3 \longleftrightarrow \{d_i\}_{i=1}^3,$$

$\mathbb{P}$ and $\{d_i\}_{i=1}^3$ are also a mathematical model of 3 coin tosses.

Readers may suspect that, in the first place, (1.1) is not correct for real coin tosses. Indeed, rigorously speaking, since Heads and Tails are differently carved, (1.1) is not exact in the real world. What we call 'coin tosses' is an idealized model, which can exist only in our mind—just as we consider the equation $(x-a)^2 + (y-b)^2 = c^2$ as a mathematical model of circle, although there is no true circle in the real world.

1.1.1 *Probability space*

Let us present what we stated in the previous section in a general setup. In what follows, 'probability theory' means the axiomatic system for probability established by [Kolomogorov (1933)]†[2] and all its derived theorems.

Let us begin with probability distribution and probability space.

Definition 1.1. (Probability distribution) Let Ω be a non-empty finite set, i.e., $\Omega \neq \emptyset$†[3] and $\#\Omega < \infty$. Suppose that for each $\omega \in \Omega$, there corresponds a real number $0 \leqq p_\omega \leqq 1$ so that

$$\sum_{\omega \in \Omega} p_\omega = 1$$

(Sec. A.1.2). Then, we call the set of all pairs ω and p_ω

$$\{(\omega, p_\omega) \mid \omega \in \Omega\}$$

a *probability distribution* (or simply, a *distribution*) in Ω.

Definition 1.2. (Probability space) Let Ω be a non-empty finite set and let $\mathfrak{P}(\Omega)$ be the power set of Ω. If a function $P : \mathfrak{P}(\Omega) \to \mathbb{R}$ satisfies
(i) $0 \leqq P(A) \leqq 1, \quad A \in \mathfrak{P}(\Omega)$,
(ii) $P(\Omega) = 1$, and

[2]See Bibliography at the end of the book.
[3]'$\emptyset$' denotes the empty set.

(iii) $A, B \in \mathfrak{P}(\Omega)$ are *disjoint*, i.e., $A \cap B = \emptyset$

$$\Longrightarrow P(A \cup B) = P(A) + P(B),$$

then the triplet $(\Omega, \mathfrak{P}(\Omega), P)$ is called a *probability space*.†[4] An element of $\mathfrak{P}(\Omega)$ (i.e., a subset of Ω) is called an *event*, in particular, Ω is called the *whole event* (or the *sample space*), and $\emptyset$ the *empty event*. A one point set $\{\omega\}$ or ω itself is called an *elementary event*. P is called a *probability measure* (or simply, a *probability*) and $P(A)$ the *probability of* A.

For a non-empty finite set Ω, to give a distribution in it and to give a probability space are equivalent. In fact, if a distribution $\{(\omega, p_\omega) \,|\, \omega \in \Omega\}$ is given, by defining a probability $P : \mathfrak{P}(\Omega) \to \mathbb{R}$ as

$$P(A) := \sum_{\omega \in A} p_\omega, \quad A \in \mathfrak{P}(\Omega),$$

a triplet $(\Omega, \mathfrak{P}(\Omega), P)$ becomes a probability space. Conversely, if a probability space $(\Omega, \mathfrak{P}(\Omega), P)$ is given, by defining

$$p_\omega := P(\{\omega\}), \quad \omega \in \Omega,$$

$\{(\omega, p_\omega) \,|\, \omega \in \Omega\}$ becomes a distribution in Ω.

A triplet $(\Omega, \mathfrak{P}(\Omega), P)$ is a probability space provided that the conditions (i)(ii)(iii) of Definition 1.2 are satisfied, no matter whether it is related to a random phenomenon or not. Thus a triplet

$$(\{0,1\}^3, \mathfrak{P}(\{0,1\}^3), P_3),$$

whose components have been defined in Example 1.1, is a probability space not because it is related to 3 coin tosses but because it satisfies all the conditions (i)(ii)(iii).

Like P_3, in general, a probability measure satisfying

$$P(A) = \frac{\#A}{\#\Omega}, \quad A \in \mathfrak{P}(\Omega),$$

is called a *uniform probability measure*. Equivalently, a distribution satisfying

$$p_\omega = \frac{1}{\#\Omega}, \quad \omega \in \Omega,$$

---

[4] In mathematics, many kinds of 'spaces' enter the stage, such as linear space, Euclidean space, topological space, Hilbert space, etc. These are sets accompanied by some structures, operations, or functions. In general, they have nothing to do with the 3-dimensional space where we live.

is called a *uniform distribution*. Setting the uniform distribution means that we assume every element of Ω is equally likely to be chosen.

By the way, in Example 1.2, you may wish to assume $[0,1)$ to be the whole event and $\mathbb{P}$ to be the probability, but since $[0,1)$ is an infinite set, it is not covered by Definition 1.2. By extending the definition of probability space, it is possible to consider an infinite set as a whole event, but to do this, we need Lebesgue's *measure theory*, which exceeds the level of this book.

Each assertion of the following proposition is easy to derive from Definition 1.2.

Proposition 1.1. *Let $(\Omega, \mathfrak{P}(\Omega), P)$ be a probability space. For $A, B \in \mathfrak{P}(\Omega)$, we have*
(i) $P(A^c) = 1 - P(A)$,[†5] *in particular, $P(\emptyset) = 0$,*
(ii) $A \subset B \implies P(A) \leqq P(B)$, and
(iii) $P(A \cup B) = P(A) + P(B) - P(A \cap B)$, in particular, $P(A \cup B) \leqq P(A) + P(B)$.

1.1.2 *Random variable*

Definition 1.3. Let $(\Omega, \mathfrak{P}(\Omega), P)$ be a probability space. We call a function $X : \Omega \to \mathbb{R}$ a *random variable*. Let $\{a_1, \ldots, a_s\} \subset \mathbb{R}$ be the set of all possible values that X can take, which is called the *range* of X, and let p_i be the probability that $X = a_i$:

$$P(\{\omega \in \Omega \,|\, X(\omega) = a_i\}) =: p_i, \quad i = 1, \ldots, s. \tag{1.4}$$

Then, we call the set of all pairs (a_i, p_i)

$$\{(a_i, p_i) \,|\, i = 1, \ldots, s\} \tag{1.5}$$

the *probability distribution* (or simply, the *distribution*) of X. Since we have

$$0 \leqq p_i \leqq 1, \quad i = 1, \ldots, s, \qquad p_1 + \cdots + p_s = 1,$$

(1.5) is a distribution in the range $\{a_1, \ldots, a_s\}$ of X. The left-hand side of (1.4) and the event inside $P(\ )$ are often abbreviated as

$$P(X = a_i) \quad \text{and} \quad \{X = a_i\},$$

respectively.

[5] $A^c := \{\omega \in \Omega \,|\, \omega \notin A\}$ is the complement of A,

For several random variables $X_1, \ldots, X_n$, let $\{a_{i1}, \ldots, a_{is_i}\} \subset \mathbb{R}$ be the range of X_i, $i = 1, \ldots, n$, and let

$$P(X_1 = a_{1j_1}, \ldots, X_n = a_{nj_n}) =: p_{j_1,\ldots,j_n}, \qquad j_1 = 1, \ldots, s_1, \quad \ldots, \quad j_n = 1, \ldots, s_n. \tag{1.6}$$

Then, the set

$$\{((a_{1j_1}, \ldots, a_{nj_n}), p_{j_1,\ldots,j_n}) \mid j_1 = 1, \ldots, s_1, \quad \ldots, \quad j_n = 1, \ldots, s_n\} \tag{1.7}$$

is called the *joint distribution* of $X_1, \ldots, X_n$. Of course, the left-hand side of (1.6) is an abbreviation of

$$P(\{\omega \in \Omega \mid X_1(\omega) = a_{1j_1}, X_2(\omega) = a_{2j_2}, \ldots, X_n(\omega) = a_{nj_n}\}).$$

Since we have

$$0 \leqq p_{j_1,\ldots,j_n} \leqq 1, \quad j_1 = 1, \ldots, s_1, \quad \ldots, \quad j_n = 1, \ldots, s_n,$$
$$\sum_{j_1=1,\ldots,s_1, \ \ldots, \ j_n=1,\ldots,s_n} p_{j_1,\ldots,j_n} = 1$$

(Sec. A.1.2), (1.7) is a distribution in the direct product of the ranges of $X_1, \ldots, X_n$ (Definition A.1)

$$\{a_{11}, \ldots, a_{1s_1}\} \times \cdots \times \{a_{n1}, \ldots, a_{ns_n}\}.$$

In contrast with joint distribution, the distribution of each individual X_i is called the *marginal distribution.*

Example 1.3. Let us look closely at joint distribution and marginal distribution in the case of $n = 2$. Suppose that the joint distribution of two random variables X_1 and X_2 is given by

$$P(X_1 = a_{1i}, X_2 = a_{2j}) = p_{ij}, \quad i = 1, \ldots, s_1, \quad j = 1, \ldots, s_2.$$

Then, their marginal distributions are computed as

$$P(X_1 = a_{1i}) = \sum_{j=1}^{s_2} P(X_1 = a_{1i}, X_2 = a_{2j}) = \sum_{j=1}^{s_2} p_{ij}, \quad i = 1, \ldots, s_1,$$

$$P(X_2 = a_{2j}) = \sum_{i=1}^{s_1} P(X_1 = a_{1i}, X_2 = a_{2j}) = \sum_{i=1}^{s_1} p_{ij}, \quad j = 1, \ldots, s_2.$$

This situation is illustrated in the following table.

X_2 \ X_1	a_{11}	$\cdots\cdots$	a_{1s_1}	marginal distribution of X_2
a_{21}	p_{11}	$\cdots\cdots$	p_{s_11}	$\sum_{i=1}^{s_1} p_{i1}$
$\vdots$	$\vdots$	$\ddots$	$\vdots$	$\vdots$
a_{2s_2}	p_{1s_2}	$\cdots\cdots$	$p_{s_1s_2}$	$\sum_{i=1}^{s_1} p_{is_2}$
marginal distribution of X_1	$\sum_{j=1}^{s_2} p_{1j}$	$\cdots\cdots$	$\sum_{j=1}^{s_2} p_{s_1j}$	$\sum_{\substack{i=1,\ldots,s_1 \\ j=1,\ldots,s_2}} p_{ij} = 1$

As you see, since they are placed in the margins of the table, they are called 'marginal' distributions.

The joint distribution uniquely determines the marginal distributions (Example 1.3), but in general the marginal distributions do not uniquely determine the joint distribution.

Example 1.4. The coordinate functions $\xi_i : \{0,1\}^3 \to \mathbb{R}$, $i = 1,2,3$, defined on the probability space $(\{0,1\}^3, \mathfrak{P}(\{0,1\}^3), P_3)$ by (1.2) are random variables. They all have a same distribution:

$$\{(0,1/2),(1,1/2)\}.$$

Their joint distribution is the uniform distribution in $\{0,1\}^3$:

$$\{((0,0,0),1/8),((0,0,1),1/8),((0,1,0),1/8),((0,1,1),1/8),$$
$$((1,0,0),1/8),((1,0,1),1/8),((1,1,0),1/8),((1,1,1),1/8)\}.$$

Example 1.5. A constant can be considered to be a random variable. Let $(\Omega, \mathfrak{P}(\Omega), P)$ be a probability space and $c \in \mathbb{R}$ be a constant. Then, a random variable $X(\omega) := c$, $\omega \in \Omega$, has a distribution $\{(c,1)\}$.

Random variables play the leading role in probability theory. "A random variable X is defined on a probability space $(\Omega, \mathfrak{P}(\Omega), P)$" is interpreted as "$\omega$ is randomly chosen from Ω with probability $P(\{\omega\})$, and accordingly the value $X(\omega)$ becomes random." In general, choosing an $\omega \in \Omega$ and getting the value $X(\omega)$ is called *sampling*, and $X(\omega)$ is called a *sample value* of X.

In probability theory, we always deal with random variables as functions, and we are indifferent to individual sample values or sampling methods. Therefore random variables need not have interpretation that they are random, and ω need not be chosen randomly. However in practical applications, such as mathematical statistics or the Monte Carlo method, sample values or sampling methods may become significant.

Usually, a probability space is just a stage that random variables enter. Given a distribution or a joint distribution, we often make a suitable probability space and define a random variable or several random variables on it, so that its distribution or their joint distribution coincides with the given one. For example, for any given distribution $\{ (a_i, p_i) \mid i = 1, \ldots, s \}$, define a probability space $(\Omega, \mathfrak{P}(\Omega), P)$ and a random variable X by

$$\Omega := \{a_1, \ldots, a_s\}, \quad P(\{a_i\}) := p_i, \quad X(a_i) := a_i, \quad i = 1, \ldots, s.$$

Then, the distribution of X coincides with the given one. Similarly, for any given joint distribution

$$\{((a_{1j_1}, \ldots, a_{nj_n}), p_{j_1,\ldots,j_n}) \mid j_1 = 1, \ldots, s_1, \ \ldots, \ j_n = 1, \ldots, s_n \}, \quad (1.8)$$

define a probability space $(\Omega, \mathfrak{P}(\Omega), P)$ and random variables $X_1, \ldots, X_n$ by

$$\begin{aligned} \Omega &:= \{a_{11}, \ldots, a_{1s_1}\} \times \cdots \times \{a_{n1}, \ldots, a_{ns_n}\}, \\ P(\{(a_{1j_1}, \ldots, a_{nj_n})\}) &:= p_{j_1,\ldots,j_n}, \quad j_1 = 1, \ldots, s_1, \ \ldots, \ j_n = 1, \ldots, s_n, \\ X_i((a_{1j_1}, \ldots, a_{nj_n})) &:= a_{ij_i}, \quad i = 1, \ldots, n. \end{aligned}$$

Each X_i is a coordinate function. Then, the joint distribution of $X_1, \ldots, X_n$ coincides with the given one (1.8). Such realization of probability space and random variable(s) is called the *canonical realization.* Example 1.4 shows the canonical realization of 3 coin tosses.

Remark 1.1. Laplace insisted that randomness should not exist and all phenomena should be deterministic ([Lapalce (1812)]). For an intelligence who knows all about the forces that move substances and about the positions and the velocities of all molecules that consist the substances at an initial time, and if in addition the intelligence†[6] has a vast ability to analyze the motion equation, there would be no irregular things in this world and everything would be deterministic. However, for us who know only a little part of the universe and do not have enough ability to analyze the very complicated motion equation, things occur as if they do at random.

[6]This intelligence is often referred to as *Laplace's demon.*

The formulation of random variable in probability theory reflects Laplace's determinism. Namely, the whole event can be interpreted as the set of all possible initial values, each elementary event ω as one of the initial values, and $X(\omega)$ as the solution of the very complicated motion equation under the initial value ω.

Laplace's determinism had been a dominating thought in the history of science, until quantum mechanics was discovered.

1.2 Random number

Randomness is involved in the procedure that an ω is chosen from Ω, on the other hand, probability theory is indifferent to that procedure. Therefore it seems impossible to analyze randomness by probability theory, but as a matter of fact, it is possible. We explain why it is possible in this and in the following sections.

Randomness can be defined in mathematics by formulating the procedure of choosing an ω from Ω.

Generalizing Example 1.1, consider

$$(\{0,1\}^n, \mathfrak{P}(\{0,1\}^n), P_n)$$

as a probability space for n coin tosses, where $P_n : \mathfrak{P}(\{0,1\}^n) \to \mathbb{R}$ is the uniform probability measure on $\{0,1\}^n$, i.e., the probability measure satisfying

$$P_n(A) := \frac{\#A}{\#\{0,1\}^n} = \frac{\#A}{2^n}, \quad A \in \mathfrak{P}(\{0,1\}^n). \tag{1.9}$$

Suppose that Alice†7 chooses an $\omega \in \{0,1\}^n$ of her own will. When n is small, she can easily write down a $\{0,1\}$-sequence of length n for ω. For example, if $n = 10$, she writes $(1,1,1,0,1,0,0,1,1,1)$. When $n = 1000$, she can do it somehow in the same way.

It comes into question when $n \gg 1$.†8 For example, when $n = 10^8$, how on earth can Alice choose an ω from $\{0,1\}^{10^8}$? In principle, she should write down a $\{0,1\}$-sequence of length 10^8, but it is impossible because 10^8 is too huge. Considering the hardness of the task, she cannot help using a computer to choose an $\omega \in \{0,1\}^{10^8}$. The computer program that produces

$$\omega = (0,0,0,0,\ldots,0,0) \in \{0,1\}^{10^8}, \tag{1.10}$$

[7] Alice is a fictitious character who makes several thought experiments in this book.
[8] $a \gg b$ means "a is much greater than b". See Sec. A.1.3.

i.e., the run of 0's of length 10^8, would be simple and easy to write. The one that produces

$$\omega = (1, 0, 1, 0, \ldots, 1, 0) \in \{0, 1\}^{10^8}, \tag{1.11}$$

i.e., 5×10^7 times repetition of a pattern '1, 0' would be less simple but still easy to write. On the other hand, for some $\omega \in \{0,1\}^{10^8}$, the program to produce it would be too long to write in practice. Let us explain it below.

In general, a program is a finite string of letters and symbols, which is described in computer as a $\{0,1\}$-sequence of finite length. For each $\omega \in \{0,1\}^{10^8}$, let q_ω be the shortest program that produces ω, and let $L(q_\omega)$ denote the length[†9] of q_ω as a $\{0,1\}$-sequence. If $\omega \neq \omega'$, then $q_\omega \neq q_{\omega'}$. Now, the number of ω for which $L(q_\omega) = k$ is at most 2^k, the number of all elements of $\{0,1\}^k$. This implies that the total number of $\omega \in \{0,1\}^{10^8}$ for which $L(q_\omega) \leqq M$ is at most

$$\begin{aligned} \#\{0,1\}^1 + \#\{0,1\}^2 + \cdots + \#\{0,1\}^M &= 2^1 + 2^2 + \cdots + 2^M \\ &= 2^{M+1} - 2. \end{aligned}$$

Conversely, the number of $\omega \in \{0,1\}^{10^8}$ for which $L(q_\omega) \geqq M+1$ is at least $2^{10^8} - 2^{M+1} + 2$. More concretely, the number of $\omega \in \{0,1\}^{10^8}$ for which $L(q_\omega) \geqq 10^8 - 10$ is at least $2^{10^8} - 2^{10^8 - 10} + 2$, which account for at least $1 - 2^{-10} = 1023/1024$ of all elements in $\{0,1\}^{10^8}$. On the other hand, for any $\omega \in \{0,1\}^{10^8}$, there is a program that outputs ω as it is—whose length would be a little greater than and hence approximately equal to the length of ω $(= 10^8)$. These facts show that for nearly all $\omega \in \{0,1\}^{10^8}$, we have $L(q_\omega) \approx 10^8$.[†10] In general, when $n \gg 1$, we call an $\omega \in \{0,1\}^n$ with $L(q_\omega) \approx n$ a *random* $\{0,1\}$*-sequence* or a *random number*.[†11]

An ω that can be produced by a short program q_ω can be done so because it has some regularity, or rather, q_ω itself describes the regularity of ω. The longer q_ω is, the less regular ω is. Therefore we can say that random numbers are the least regular $\{0,1\}$-sequences. Random numbers require too long programs to produce them. This means that they cannot be chosen by Alice. Thus the notion of random number well expresses randomness that we intuitively know.

---

[9] The length of the shortest program depends on the programming language, but do not mind it here. We will explain it in detail in Sec. 2.2.1.

[10] $x \approx y$ means "x and y are approximately equal to each other".

[11] In this manner, the definition of random number cannot help being somewhat ambiguous. We will explain it in detail in Remark 2.1.

When $n \gg 1$, if each $\omega \in \{0,1\}^n$ is chosen with probability $P_n(\{\omega\}) = 2^{-n}$, a random number will be chosen with very high probability. At the opening of this chapter, we mentioned "Tossing a coin many times, record 1 if it comes up Heads and record 0 if it comes up Tails at each coin toss. Then, we get a long $\{0,1\}$-sequence that is random." Exactly speaking, we must revise it as "we get a long random $\{0,1\}$-sequence with very high probability." Indeed, with very small probability, we may have a non-random $\{0,1\}$-sequence such as (1.10) or (1.11).

Remark 1.2. In probability theory, the term 'random' usually means that the quantity in question is a random variable. In general sciences, 'random' is used in many contexts with various meanings. The randomness introduced in this section is called *algorithmic randomness* to specify the meaning.

Remark 1.3. We have defined randomness for long $\{0,1\}$-sequences, such as the outcomes of many coin tosses, but we feel even a 'single coin toss' is random. Why do we feel so?

For example, if we drop a coin with Heads up quietly from a height of 5mm above a desk surface, then the outcome will be Heads without doubt. It is thus possible to control Heads and Tails in this case. On the other hand, if we drop it from a height of 50cm, it is not possible. Since rebounding obeys a deterministic physical law, we can, in principle, control Heads and Tails, even though the coin is dropped from an arbitrarily high position (Remark 1.1). However, each rebounding motion depends so acutely on the initial value that a slight difference of initial values causes a big difference in the results (Fig. 1.2). In other words, when we drop a coin from a height of 50cm, in order to control its Heads and Tails, we must measure and set the initial mechanical state of the coin with extremely high precision, which is impossible in practice. Thus controlling the Heads and Tails of a coin dropped from a high position is similar to choosing a random number in that both are beyond our ability.

1.3 Limit theorem

1.3.1 *Analysis of randomness*

Since mathematical models of virtually all random phenomena can be constructed from coin tosses (Sec. 1.5.2), to analyze randomness we have essentially only to study properties of random numbers. However, we cannot

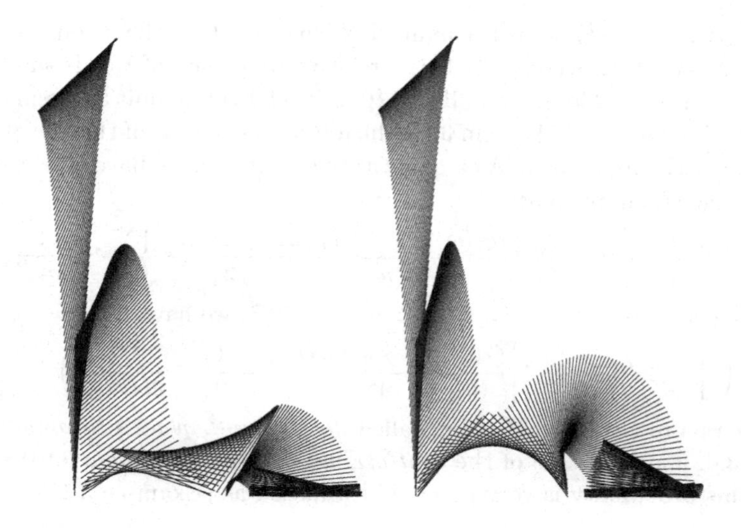

Fig. 1.2 Reboundings of a coin (Two simulations with slightly different initial values)

know any property of each individual random number. Indeed, if ω is a random number, we cannot know that it is so (Theorem 2.7). What is sure is that when $n \gg 1$, random numbers account for nearly all $\{0,1\}$-sequences. In such a situation, is it possible to study properties of random numbers?

To study properties of them *as a whole* is possible. The answer is unexpectedly simple: study properties that nearly all long $\{0,1\}$-sequences share. Such properties are stated by various *limit theorems* in probability theory. Therefore the study of limit theorems is the most important in probability theory.

Here is an example of limit theorem. For each $i = 1, \ldots, n$, being defined on the probability space $(\{0,1\}^n, \mathfrak{P}(\{0,1\}^n), P_n)$, the coordinate function† [12]

$$\xi_i(\omega) := \omega_i, \quad \omega = (\omega_1, \ldots, \omega_n) \in \{0,1\}^n, \tag{1.12}$$

represents the i-th outcome (0 or 1) of n coin tosses. Then, no matter how small $\varepsilon > 0$ is, it holds that

$$\lim_{n\to\infty} P_n\left(\left\{\omega \in \{0,1\}^n \;\middle|\; \left|\frac{\xi_1(\omega) + \cdots + \xi_n(\omega)}{n} - \frac{1}{2}\right| > \varepsilon\right\}\right) = 0. \tag{1.13}$$

[12] Strictly speaking, since the domain of definition of ξ_i is $\{0,1\}^n$, which depends on n, we must write it as $\xi_{n,i}$. We drop n because its meaning is common to any n: the i-th coordinate of ω.

Since $\xi_1(\omega)+\cdots+\xi_n(\omega)$ is the number of Heads(= 1) in the n coin tosses ω, (1.13) asserts that when $n \gg 1$, the relative frequency of Heads should be approximately 1/2 for nearly all $\omega \in \{0,1\}^n$. (1.13) is a limit theorem called *Bernoulli's theorem* (Theorem 3.2, which is a special case of the *law of large numbers* (Theorem 3.3)). As a quantitative estimate, we have Chebyshev's inequality (Example 3.3):

$$P_n\left(\left\{\omega \in \{0,1\}^n \,\middle|\, \left|\frac{\xi_1(\omega)+\cdots+\xi_n(\omega)}{n} - \frac{1}{2}\right| \geqq \varepsilon\right\}\right) \leqq \frac{1}{4n\varepsilon^2}.$$

As a special case where $\varepsilon = 1/2000$ and $n = 10^8$, we have

$$P_{10^8}\left(\left\{\omega \in \{0,1\}^{10^8} \,\middle|\, \left|\frac{\xi_1(\omega)+\cdots+\xi_{10^8}(\omega)}{10^8} - \frac{1}{2}\right| \geqq \frac{1}{2000}\right\}\right) \leqq \frac{1}{100}.$$

A much more advanced theorem called *de Moivre–Laplace's theorem* (Theorem 3.6, a special case of the *central limit theorem* (Theorem 3.9)) shows that this probability is very precisely estimated as (Example 3.10)†[13]

$$P_{10^8}\left(\left\{\omega \in \{0,1\}^{10^8} \,\middle|\, \left|\frac{\xi_1(\omega)+\cdots+\xi_{10^8}(\omega)}{10^8} - \frac{1}{2}\right| \geqq \frac{1}{2000}\right\}\right)$$
$$\approx 2\int_{9.9999}^{\infty} \frac{1}{\sqrt{2\pi}} \exp\left(-\frac{x^2}{2}\right) dx = 1.52551 \times 10^{-23}.$$

Subtract both sides from 1, we get

$$P_{10^8}\left(\left\{\omega \in \{0,1\}^{10^8} \,\middle|\, \left|\frac{\xi_1(\omega)+\cdots+\xi_{10^8}(\omega)}{10^8} - \frac{1}{2}\right| < \frac{1}{2000}\right\}\right)$$
$$\approx 1 - 1.52551 \times 10^{-23}.$$

Namely, $1 - 1.52551 \times 10^{-23}$ of all the elements of $\{0,1\}^{10^8}$ are those ω that satisfy

$$\left|\frac{\xi_1(\omega)+\cdots+\xi_{10^8}(\omega)}{10^8} - \frac{1}{2}\right| < \frac{1}{2000}. \tag{1.14}$$

On the other hand, nearly all elements of $\{0,1\}^{10^8}$ are random numbers. Therefore nearly all random numbers ω share the property (1.14), and conversely, nearly all ω that satisfy (1.14) are random numbers.

Such as ω in (1.11), there are $\omega \in \{0,1\}^{10^8}$ that are not random but satisfy (1.14). In general, an event of probability close to 1 that a limit theorem specifies does not completely coincide with the set of random numbers, but the difference between them is very little (Fig. 1.3).

---

[13] $\exp(x)$ is an alternative description for the exponential function e^x, hence $\exp(-x^2/2)$ stands for $e^{-x^2/2}$, and $\int_{9.9999}^{\infty}$ is an abbreviation of $\lim_{R\to\infty}\int_{9.9999}^{R}$, which is called an *improper integral.*

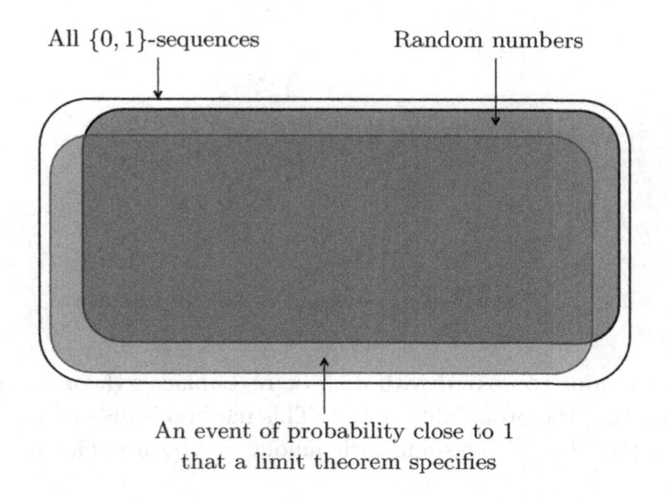

Fig. 1.3 Random number and limit theorem (Conceptual figure)

Bernoulli's theorem was published in 1713 after his death. Since then, limit theorems have been the central theme of probability theory. On the other hand, the concept of randomness was established in 1960's. Thus mathematicians have been studying limit theorems since long before they defined randomness.

1.3.2 *Mathematical statistics*

Applying the law of large numbers and the central limit theorem, we can estimate various probabilities in the real world.

For example, a thumbtack (drawing pin) can land either point up or down. When we toss it, what is the probability it lands point up? To answer this, toss it many times, count the number of landings with point up, and calculate the relative frequency of landings with point up, which will be an approximate value of the probability. A similar method applies to opinion polls: without interviewing all citizens, choose a part of citizens by dice or lot (*random sampling*), interview them, and compute the proportion of those who say 'yea' among them.

Suppose we are given a certain coin, and asked "When it is tossed, is the probability that it comes up Heads 1/2?" To answer this, toss it many times, calculate the relative frequency of Heads, and examine if it is

Fig. 1.4 Thumbtacks (Left: point up, Right: point down)

a reasonable value compared with de Moivre–Laplace's theorem under the assumption that the probability is $1/2$. This method is also useful when we examine if the ratio of boys and girls among newborn babies in a certain country is $1:1$.

In this way, through experiments, observations, and investigations, making a mathematical model for random phenomena, testing it, or predicting the random phenomena by it, are all included in a branch of study called *mathematical statistics*.

1.4 Monte Carlo method

The Monte Carlo method is a numerical method to solve mathematical problems by computer-aided sampling of random variables.

When $\#\Omega$ is small, sampling of a random variable $X : \Omega \to \mathbb{R}$ is easy. If $\#\Omega = 10^8$, we have only to specify a number of at most 9 decimal digits to choose an $\omega \in \Omega$, but when $\Omega = \{0,1\}^{10^8}$, a computer is indispensable for sampling. Let us consider the following exercise.

> **Exercise I** When we toss a coin 100 times, what is the probability p that it comes up Heads at least 6 times in succession?

We apply one of the ideas of mathematical statistics stated in Sec. 1.3.2. Repeat '100 coin tosses' 10^6 times, and let S be the number of the occurrences of "the coin comes up Heads 6 times in succession" among the 10^6 trials. Then, by the law of large numbers, $S/10^6$ will be a good approximate value of p with high probability. To do this, the total number of coin tosses we need is $100 \times 10^6 = 10^8$. Of course, we do not toss a real coin so many times, instead we use a computer.

Let us sort the problem. S is formulated as a random variable defined on the probability space $(\{0,1\}^{10^8}, \mathfrak{P}(\{0,1\}^{10^8}), P_{10^8})$. In this case, Chebyshev's inequality shows

$$P_{10^8}\left(\left|\frac{S(\omega)}{10^6} - p\right| \geqq \frac{1}{200}\right) \leqq \frac{1}{100}$$

(Example 4.2). Subtract the both sides from 1, we have

$$P_{10^8}\left(\left|\frac{S(\omega)}{10^6} - p\right| < \frac{1}{200}\right) \geqq \frac{99}{100}. \tag{1.15}$$

Namely, if Alice chooses an ω from $\{0,1\}^{10^8}$, and computes the value of $S(\omega)/10^6$, she can get an approximate value of p with error less than $1/200$ with probability at least 0.99.

Now, to give the inequality (1.15) a practical meaning, Alice should be equally likely to choose an $\omega \in \{0,1\}^{10^8}$. This means that she should choose ω from mainly among random numbers because they account for nearly all elements of $\{0,1\}^{10^8}$.†14 However, as we saw in Sec. 1.2, it is impossible to choose a random number even by computer.

In most of practical Monte Carlo methods, pseudorandom numbers are used instead of random numbers. A program that produces pseudorandom numbers, mathematically speaking, a function

$$g : \{0,1\}^l \to \{0,1\}^n, \quad l < n,$$

is called a *pseudorandom generator*. For practical use, l is assumed to be small enough for Alice to be equally likely to choose $\omega' \in \{0,1\}^l$, and on the other hand, n is assumed to be too large for her to be equally likely to choose $\omega \in \{0,1\}^n$. Namely, $l \ll n$. The program produces $g(\omega') \in \{0,1\}^n$ from $\omega' \in \{0,1\}^l$ that Alice has chosen. Here, $g(\omega')$ is called a *pseudorandom number*, and ω' is called its *seed*.

For any ω', the pseudorandom number $g(\omega')$ is not a random number. Nevertheless, it is useful in some situations. In fact, in the case of Exercise I, there exists a suitable pseudorandom generator $g : \{0,1\}^{238} \to \{0,1\}^{10^8}$ such that an inequality

$$P_{238}\left(\left|\frac{S(g(\omega'))}{10^6} - p\right| < \frac{1}{200}\right) \geqq \frac{99}{100},$$

which is similar to (1.15), holds (Fig. 1.5, the *Random Weyl sampling* (Example 4.4)).

---

[14]This is the reason why random number is said to be needed for the Monte Carlo method.

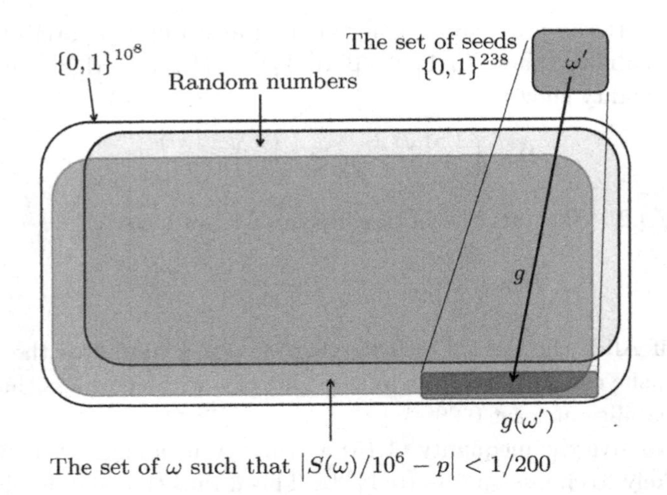

Fig. 1.5 The role of $g : \{0,1\}^{238} \to \{0,1\}^{10^8}$ (Conceptual figure)

Alice can choose any $\omega' \in \{0,1\}^{238}$ of her own will, or by tossing a real coin 238 times, she can get a seed ω' by random sampling. Thus to solve Exercise I, she does not need a random number, but by the pseudorandom number that g produces from her seed, she can actually get a good approximate value of p with high probability.

Exercise I deals with a problem about coin tosses, but practical problems to which the Monte Carlo method is applied are much more complicated. Nevertheless, since any practical probabilistic problem can be reduced to that of coin tosses (Sec. 1.5.2), we may assume pseudorandom numbers to be $\{0,1\}$-sequences.

1.5 Infinite coin tosses

Borel's model of coin tosses (Example 1.2) can give a mathematical model of not only 3 coin tosses but also arbitrarily many coin tosses. Furthermore, the sequence of functions $\{d_i\}_{i=1}^{\infty}$ defined in Example 1.2 can be regarded as *infinite* coin tosses. Of course, there do not exist infinite coin tosses in the real world, but by some reasons, it is important to consider them.

The contents of this section slightly exceeds the level of this book.

1.5.1 *Borel's normal number theorem*

Rational numbers are sufficient for practical computation, but to let calculus be available, real numbers are necessary. Just like this, the first reason why we consider infinite coin tosses is that they are useful when we analyze the limit behavior of n coin tosses as $n \to \infty$. Indeed, the fact that the probability space for n coin tosses varies as n varies is awkward, bad-looking, and deficient for advanced study of probability theory.

For example, Borel's *normal number theorem*

$$\mathbb{P}\left(\lim_{n\to\infty} \frac{1}{n} \sum_{i=1}^{n} d_i = \frac{1}{2} \right) = 1 \tag{1.16}$$

asserts just one of the analytic properties of the sequence of functions $\{d_i\}_{i=1}^{\infty}$, but it is interpreted in the context of probability theory as "When we toss a coin infinitely many times, the asymptotic limit of the relative frequency of Heads is 1/2 with probability 1." It is known that Borel's normal number theorem implies Bernoulli's theorem (1.13). Note that it is not easy to grasp the exact meaning of (1.16). Intuitively, it means that the 'length' of the set

$$A := \left\{ x \in [0,1) \;\middle|\; \lim_{n\to\infty} \frac{1}{n} \sum_{i=1}^{n} d_i(x) = \frac{1}{2} \right\} \subset [0,1)$$

is equal to 1, but since A is not a simple set like a semi-open interval, how to define its 'length' and how to compute it come into question. To solve them, we need measure theory.

1.5.2 *Construction of Brownian motion*

The second reason why we consider infinite coin tosses is that we can construct a random variable with arbitrary distribution from infinite coin tosses. What is more, except very special cases†15, any probabilistic object can be constructed from them. As an example, we here construct from them a *Brownian motion*—the most important stochastic process both in theory and in practice.

Define a function $F : \mathbb{R} \to (0,1) := \{ x \,|\, 0 < x < 1 \} \subset \mathbb{R}$ as follows.†16

$$F(t) := \int_{-\infty}^{t} \frac{1}{\sqrt{2\pi}} \exp\left(-\frac{u^2}{2} \right) du, \quad t \in \mathbb{R}.$$

[15] For example, construction of uncountable (cf. p.25) independent random variables.

[16] $\int_{-\infty}^{t}$ is an abbreviation of $\lim_{R\to\infty} \int_{-R}^{t}$, which is also called an improper integral. See Remark 3.7.

The integrand in the right-hand side is the probability density function of the *standard normal distribution* (or the *standard Gaussian distribution*). Since $F : \mathbb{R} \to (0,1)$ is a continuous increasing function, its inverse function $F^{-1} : (0,1) \to \mathbb{R}$ exists. Then, putting

$$X(x) := \begin{cases} F^{-1}(x) & (0 < x < 1), \\ -\infty & (x = 0), \end{cases}$$

it holds that

$$\begin{aligned} \mathbb{P}(X < t) &:= \mathbb{P}(\{x \in [0,1) \mid X(x) < t\}) \\ &= \mathbb{P}(\{x \in [0,1) \mid x < F(t)\}) \\ &= \mathbb{P}(\,[0, F(t))\,) = F(t), \quad t \in \mathbb{R}. \end{aligned}$$

In the context of probability theory, X is interpreted as a 'random variable'.†[17] Accordingly, the above expression shows that "The probability that $X < t$ is $F(t)$", in other words, "X obeys the standard normal distribution".

Now, since $x = \sum_{i=1}^{\infty} 2^{-i} d_i(x)$, it is clear that

$$\mathbb{P}\left(\left\{ x \in [0,1) \;\middle|\; \sum_{i=1}^{\infty} 2^{-i} d_i(x) < t \right\}\right) = \mathbb{P}(\,[0,t)\,) = t, \quad t \in [0,1).$$

Any subsequence $\{d_{i_j}\}_{j=1}^{\infty}$, $1 \leqq i_1 < i_2 < \cdots$, is also infinite coin tosses, and hence we have

$$\mathbb{P}\left(\left\{ x \in [0,1) \;\middle|\; \sum_{j=1}^{\infty} 2^{-j} d_{i_j}(x) < t \right\}\right) = t, \quad t \in [0,1).$$

Accordingly, we see

$$\mathbb{P}\left(\left\{ x \in [0,1) \;\middle|\; X\left(\sum_{j=1}^{\infty} 2^{-j} d_{i_j}(x)\right) < t \right\}\right) = F(t), \quad t \in \mathbb{R}.$$

Namely, $X\left(\sum_{j=1}^{\infty} 2^{-j} d_{i_j}\right)$ obeys the standard normal distribution.

[17] We want to regard $[0,1)$ as a whole event, $\mathbb{P}$ as a probability measure, but $[0,1)$ is an infinite set and hence this idea exceeds the level of this book. Here the quotation mark of 'random variable' shows that this word is not rigorously defined in this book. In what follows, 'independent' will be used similarly.

Here is an amazing idea: if we put

$$\begin{aligned} X_1 &:= X\left(2^{-1}d_1 + 2^{-2}d_3 + 2^{-3}d_6 + 2^{-4}d_{10} + 2^{-5}d_{15} + \cdots\right), \\ X_2 &:= X\left(2^{-1}d_2 + 2^{-2}d_5 + 2^{-3}d_9 + 2^{-4}d_{14} + \cdots\right), \\ X_3 &:= X\left(2^{-1}d_4 + 2^{-2}d_8 + 2^{-3}d_{13} + \cdots\right), \\ X_4 &:= X\left(2^{-1}d_7 + 2^{-2}d_{12} + \cdots\right), \\ X_5 &:= X\left(2^{-1}d_{11} + \cdots\right), \\ &\vdots \end{aligned}$$

then each X_n obeys the standard normal distribution. We emphasize that each d_k appears only in one X_n, which means the value of each X_n does not make influence on any other $X_{n'}$ $(n' \neq n)$. Namely, $\{X_n\}_{n=1}^{\infty}$ are 'independent'.

Now, we are at the position to define a Brownian motion $\{B_t\}_{0 \leqq t \leqq \pi}$ (Fig. 1.6):

$$B_t := \frac{t}{\sqrt{\pi}} X_1 + \sqrt{\frac{2}{\pi}} \sum_{n=1}^{\infty} \frac{\sin nt}{n} X_{n+1}, \quad 0 \leqq t \leqq \pi. \tag{1.17}$$

To tell the truth, the graph of Fig. 1.6 is not exactly the Brownian motion (1.17) itself, but its approximation†[18] $\{\hat{B}_t\}_{0 \leqq t \leqq \pi}$. Let $\{X_n\}_{n=1}^{1000}$ be approximated by $\{\hat{X}_n\}_{n=1}^{1000}$, where

$$\begin{aligned} \hat{X}_1 &:= X\left(2^{-1}d_1 + 2^{-2}d_2 + 2^{-3}d_3 + \cdots + 2^{-31}d_{31}\right), \\ \hat{X}_2 &:= X\left(2^{-1}d_{32} + 2^{-2}d_{33} + 2^{-3}d_{34} + \cdots + 2^{-31}d_{62}\right), \\ \hat{X}_3 &:= X\left(2^{-1}d_{63} + 2^{-2}d_{64} + 2^{-3}d_{65} + \cdots + 2^{-31}d_{93}\right), \\ \hat{X}_4 &:= X\left(2^{-1}d_{94} + 2^{-2}d_{95} + 2^{-3}d_{96} + \cdots + 2^{-31}d_{124}\right), \\ &\vdots \\ \hat{X}_{1000} &:= X\left(2^{-1}d_{30970} + 2^{-2}d_{30971} + 2^{-3}d_{30972} + \cdots + 2^{-31}d_{31000}\right), \end{aligned}$$

and using these, define $\{\hat{B}_t\}_{0 \leqq t \leqq \pi}$ by

$$\hat{B}_t := \frac{t}{\sqrt{\pi}} \hat{X}_1 + \sqrt{\frac{2}{\pi}} \sum_{n=1}^{999} \frac{\sin nt}{n} \hat{X}_{n+1}, \quad 0 \leqq t \leqq \pi. \tag{1.18}$$

Thus, Fig. 1.6 is based on (1.18) with the sample of 31,000 coin tosses $\{d_i\}_{i=1}^{31000}$.†[19]

---

[18]To be precise, approximation in the sense of distribution.

[19]The sample of 31,000 coin tosses is produced by a pseudorandom generator in [Sugita (2011)] Sec. 4.2.

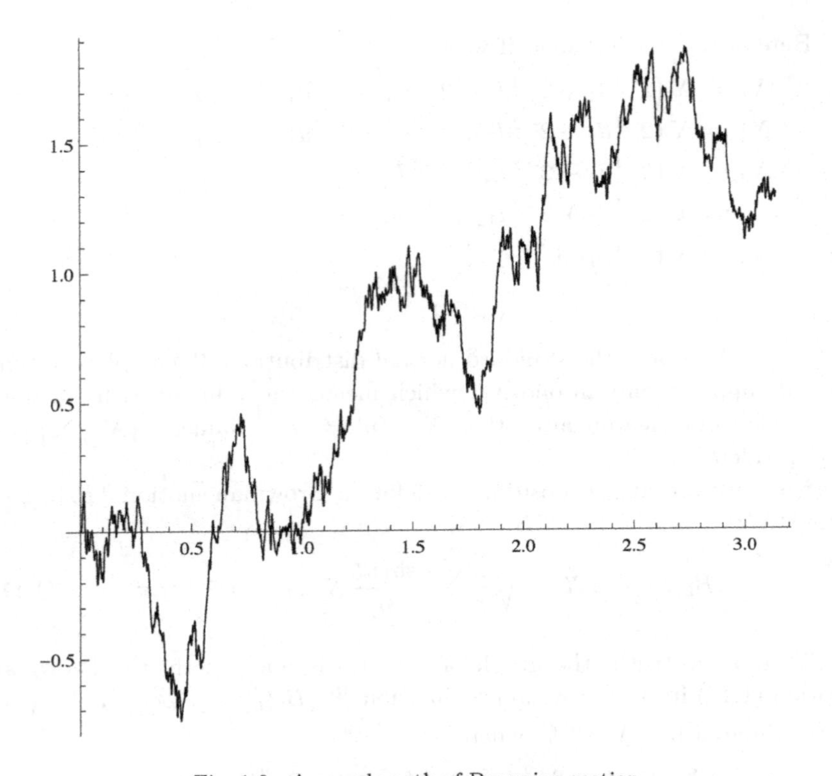

Fig. 1.6 A sample path of Brownian motion

We have constructed a Brownian motion from the infinite coin tosses $\{d_i\}_{i=1}^{\infty}$. Do not be surprised yet. If we apply the same method to infinite coin tosses $d_1, d_3, d_6, d_{10}, d_{15}, \ldots$, which compose X_1, we can construct a Brownian motion from them. Similarly, from the infinite coin tosses $d_2, d_5, d_9, d_{14}, \ldots$, which compose X_2, we can construct another 'independent' Brownian motion. Repeating this procedure, we can construct infinite 'independent' Brownian motions from $\{d_i\}_{i=1}^{\infty}$.

Chapter 2

Random number

What is randomness? To this question, which puzzled many scholars of all ages and countries, many answers were presented. For example, as we saw in Remark 1.1, Laplace stated a determinist view in his book [Lapalce (1812)]. By Kolmogorov's axiomatization of probability theory ([Kolomogorov (1933)]), a random variable was formulated as a function $X : \Omega \to \mathbb{R}$, and individual sample values $X(\omega)$ as well as sampling methods were passed over in silence. Thanks to it, mathematicians were released from the question.

In 1950's, the situation has changed since computer-aided sampling of random variables was realized: the Monte Carlo method came into the world. Before then, sampling of random variables had been done in mathematical statistics, for which *tables of random numbers*—arrays of numbers obtained by rolling dice, etc.—were used. However, they were useless for large-scale sampling by computer. Consequently, people had to consider how to make a huge table of random numbers by computer. Then, "*What is randomness?*" again came into question.

In order to define randomness, we have to formulate the procedure of choosing an ω from Ω. For this purpose, maturation of *computation theory* was indispensable. Finally, in 1960's, the definition of randomness was declared by Kolmogorov, Chaitin, and Solomonoff, independently.

This chapter deals with computation theory. Rigorous treatment of the theory needs considerably many preliminaries, and hence exceeds the level of this book. Instead of giving detailed proofs of theorems, we explain the meanings of the theorems by comparing them with mechanisms of computer.

2.1 Recursive function

The kinds of data dealt with by computer are diverse. As for input, data from keyboard, mouse, scanner, and video camera, and as for output, document, image, sound, movie, control signal for IT device, etc. They all are converted into finite $\{0, 1\}$-sequences, and then they are recorded in computer memory or disks (Fig. 2.1), copied, or transmitted.

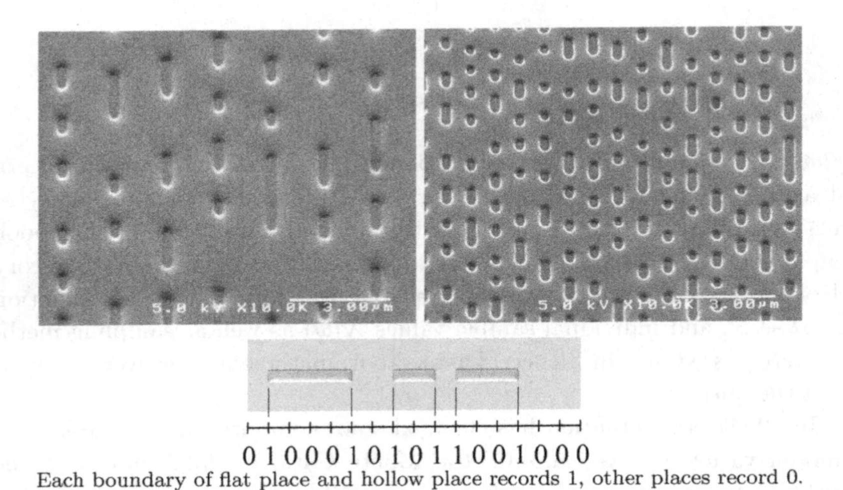

Each boundary of flat place and hollow place records 1, other places record 0.

Fig. 2.1 Images of CD (left) and DVD (right) by scanning electron microscope[†1]

Since each finite $\{0, 1\}$-sequence corresponds to a non-negative integer through the binary numeral system (Definition 2.3), every input/output data dealt with by computer can be regarded as a non-negative integer. Thus every action of computer can be regarded as a function $f : \mathbb{N} \to \mathbb{N}$.[†2] It may sound surprising that in the final analysis, computer can calculate only non-negative integers. Such extreme simplification seems to be useless, but as a matter of fact, it often leads to a great development in mathematics.

Now, conversely, can an arbitrary function $f : \mathbb{N} \to \mathbb{N}$ be realized by an action of computer? The answer is '*No*'. Even computer is not almighty. In this section, we introduce the class of 'recursive functions', which was presented as computable functions, and their basic properties.

[1] Source: Japan Science and Technology Agency 'Rika Network'

[2] $\mathbb{N} := \{0, 1, 2, \ldots\}$ is the set of all non-negative integers.

2.1.1 *Computable function*

Each action of computer is described by a program, which just like any other input/output data is a finite $\{0,1\}$-sequence, or a non-negative integer. Therefore, by numbering all programs, we can make a one-to-one correspondence between all computable functions and all non-negative integers $\mathbb{N}$. In general, if there is a one-to-one correspondence between a set A and $\mathbb{N}$, the set A is called a *countable* set, and if there is not, it is called an *uncountable* set.

Proposition 2.1. *The set* $\{\, f \mid f : \mathbb{N} \to \{0,1\}\,\}$, *i.e., the set of all functions from* $\mathbb{N}$ *to* $\{0,1\}$, *is an uncountable set.*

Proof. We show the proposition by contradiction. Suppose $\{\, f \mid f : \mathbb{N} \to \{0,1\}\,\}$ is a countable set and numbered as $\{f_0, f_1, \ldots\}$. Then, if we define a function $g : \mathbb{N} \to \{0,1\}$ by

$$g(n) := 1 - f_n(n), \quad n \in \mathbb{N},$$

we have $g(n) \neq f_n(n)$ for every $n \in \mathbb{N}$, i.e., g does not belong to the numbered set $\{f_0, f_1, \ldots\}$. This is a contradiction. Therefore $\{\, f \mid f : \mathbb{N} \to \{0,1\}\,\}$ is an uncountable set.

$$\begin{array}{cccccc}
f_0(0) & f_0(1) & f_0(2) & f_0(3) & f_0(4) & \cdots \\
f_1(0) & f_1(1) & f_1(2) & f_1(3) & f_1(4) & \cdots \\
f_2(0) & f_2(1) & f_2(2) & f_2(3) & f_2(4) & \cdots \\
f_3(0) & f_3(1) & f_3(2) & f_3(3) & f_3(4) & \cdots \\
f_4(0) & f_4(1) & f_4(2) & f_4(3) & f_4(4) & \cdots \\
\vdots & \vdots & \vdots & \vdots & \vdots & \ddots
\end{array}$$

□

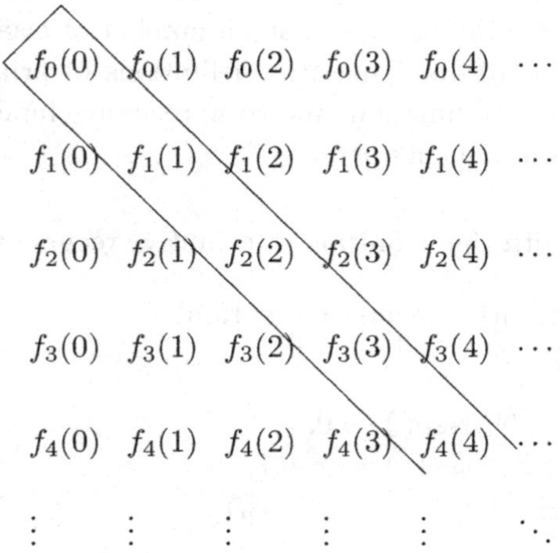

The above proof is called the *diagonal method*. The most familiar uncountable set is the set of all real numbers $\mathbb{R}$, which fact and whose proof can be found in most of textbooks of analysis or set theory.

The set of all functions $f : \mathbb{N} \to \mathbb{N}$ includes the uncountable set $\{ f \mid f : \mathbb{N} \to \{0,1\} \}$, and hence it is uncountable. The set of all computable functions is a countable subset of it, say, $\{\varphi_0, \varphi_1, \varphi_2, \dots\}$. Then, the set of all incomputable functions is uncountable because if it is countable and is numbered as $\{g_0, g_1, g_2, \dots\}$, we get

$$\{g_0, \varphi_0, g_1, \varphi_1, g_2, \varphi_2, \dots\}$$

as a numbering of the set $\{ f \mid f : \mathbb{N} \to \mathbb{N} \}$, which is a contradiction.

An important example of incomputable function is the Kolmogorov complexity (Definition 2.6, Theorem 2.7).

2.1.2 *Primitive recursive function and partial recursive function*

About the definition of computable function, there had been many discussions, until it reached a consensus in 1930's: we can compute *recursive functions* (more precisely, primitive recursive functions, partial recursive functions, and total recursive functions) and nothing else. The set of all recursive functions coincide with the set of all functions that the *Turing machine*[†3] can compute. Any actions of real computers can be described by recursive functions. It is amazing that all of diverse complicated actions of computers are just combinations of small number of basic operations.

In this subsection, we introduce the definitions of primitive recursive function, partial recursive function, and total recursive function, but we do not develop rigorous arguments here.

First, we begin with the definition of primitive recursive function.

Definition 2.1. (Primitive recursive function)
(i) (Basic functions)

$$\begin{aligned} \text{zero} &: \mathbb{N}^0 \to \mathbb{N}, \ \text{zero}(\,) := 0, \\ \text{suc} &: \mathbb{N} \to \mathbb{N}, \ \ \text{suc}(x) := x+1, \\ \text{proj}_i^n &: \mathbb{N}^n \to \mathbb{N}, \ \text{proj}_i^n(x_1, \dots, x_n) := x_i, \qquad i = 1, \dots, n \end{aligned}$$

are basic functions.[†4]

---

[3] A virtual computer with infinite memory. For details, see [Sipser (2012)].

[4] 'zero( )' is a constant function that returns 0. We use $\mathbb{N}^0$ as formal notation, but do not mind it. 'suc' denotes successor and 'proj_i^n' is a coordinate function, an abbreviation of projection.

(ii) (Composition)
For $g : \mathbb{N}^m \to \mathbb{N}$ and $g_j : \mathbb{N}^n \to \mathbb{N}$, $j = 1, \ldots, m$, we define $f : \mathbb{N}^n \to \mathbb{N}$ by

$$f(x_1, \ldots, x_n) := g(g_1(x_1, \ldots, x_n), \ldots, g_m(x_1, \ldots, x_n)).$$

This operation is called composition.
(iii) (Recursion)
For $g : \mathbb{N}^n \to \mathbb{N}$ and $\varphi : \mathbb{N}^{n+2} \to \mathbb{N}$, we define $f : \mathbb{N}^{n+1} \to \mathbb{N}$ by

$$\begin{cases} f(x_1, \ldots, x_n, 0) & := g(x_1, \ldots, x_n), \\ f(x_1, \ldots, x_n, y+1) & := \varphi(x_1, \ldots, x_n, y, f(x_1, \ldots, x_n, y)). \end{cases}$$

This operation is called recursion.
(iv) A function $\mathbb{N}^n \to \mathbb{N}$ is called a *primitive recursive function* if and only if it is a basic function or a function obtained from basic functions by applying finite combinations of composition and recursion.

Example 2.1. Two variables' sum $\mathrm{add}(x, y) = x+y$ is a primitive recursive function. Indeed, it is defined by

$$\begin{cases} \mathrm{add}(x, 0) & := \mathrm{proj}\,^1_1(x) = x, \\ \mathrm{add}(x, y+1) & := \mathrm{proj}\,^3_3(x, y, \mathrm{suc}(\mathrm{add}(x, y))). \end{cases}$$

Two variables' product $\mathrm{mult}(x, y) = xy$ is also a primitive recursive function. Indeed, it is defined by

$$\begin{cases} \mathrm{mult}(x, 0) & := \mathrm{proj}\,^2_2(x, \mathrm{zero}(\,)) = 0, \\ \mathrm{mult}(x, y+1) & := \mathrm{add}(\mathrm{proj}\,^2_1(x, y), \mathrm{mult}(x, y)). \end{cases}$$

Using a primitive recursive function $\mathrm{pred}(x) = \max\{x-1, 0\}$[†5]:

$$\begin{cases} \mathrm{pred}(0) & := \mathrm{zero}(\,) = 0, \\ \mathrm{pred}(y+1) & := \mathrm{proj}\,^2_1(y, \mathrm{pred}(y)), \end{cases}$$

we define two variables' difference $\mathrm{sub}(x, y) = \max\{x - y, 0\}$ by

$$\begin{cases} \mathrm{sub}(x, 0) & := \mathrm{proj}\,^1_1(x) = x, \\ \mathrm{sub}(x, y+1) & := \mathrm{pred}(\mathrm{proj}\,^3_3(x, y, \mathrm{sub}(x, y))), \end{cases}$$

then it is a primitive recursive function.

Secondly, we introduce partial recursive function. A *partial function* is a function g defined on a certain subset D of $\mathbb{N}^n$ and taking values in $\mathbb{N}$. We do not explicitly write the domain of definition D of g, but write it simply as $g : \mathbb{N}^n \to \mathbb{N}$.[†6] A function defined on the whole set $\mathbb{N}^n$ is called a *total function*. Since $\mathbb{N}$ is a subset of $\mathbb{N}$ itself, a total function is a partial function.

[5] 'pred' is an abbreviation of 'predecessor'. $\max\{a, b\}$ denotes the greater of a and b.
[6] This notation applies to this chapter only. In other chapters, if we write $f : E \to F$, the function f is defined for all elements of E.

Definition 2.2. (Partial recursive function)
(i) (Minimization, μ-operator)
For a partial function $p : \mathbb{N}^{n+1} \to \mathbb{N}$, we define $\mu_y(p(\bullet, \cdots, \bullet, y)) : \mathbb{N}^n \to \mathbb{N}$ by

$$\mu_y(p(x_1, \ldots, x_n, y)) := \begin{cases} \min A_p(x_1, \ldots, x_n) & (A_p(x_1, \ldots, x_n) \neq \emptyset), \\ \text{not defined} & (A_p(x_1, \ldots, x_n) = \emptyset). \end{cases}$$

Here $\min A_p(x_1, \ldots, x_n)$ is the minimum value of the following set.

$$A_p(x_1, \ldots, x_n) := \left\{ y \in \mathbb{N} \,\middle|\, \begin{array}{l} p(x_1, \ldots, x_n, z) \text{ is defined for } z \text{ such that} \\ 0 \leqq z \leqq y, \text{ and } p(x_1, \ldots, x_n, y) = 0 \end{array} \right\}.$$

(ii) A function $\mathbb{N}^n \to \mathbb{N}$ is called a *partial recursive function* if and only if it is a basic function or a partial function obtained from basic functions by applying finite combinations of composition, recursion, and minimization.

The computer programming counterpart of μ-operator is called *loop*, which is shown below.

$\mu_y(p(x_1, \ldots, x_n, y))$

(1) $y := 0$.
(2) If $p(x_1, \ldots, x_n, y) = 0$, then output y and halt.
(3) Increase y by 1, and go to (2).

The loop does not halt if $A_p(x_1, \ldots, x_n) = \emptyset$, which is called an *infinite loop*.

If a partial recursive function f does not include μ-operator (and hence it is a primitive recursive function), or if it includes $\mu_y(p(x_1, \ldots, x_n, y))$ only when $A_p(x_1, \ldots, x_n) \neq \emptyset$, f is a total function. Such f is called a *total recursive function*.

Example 2.2. The following partial recursive function

$$f(x) := \mu_y(\text{add}(\text{sub}(\text{mult}(y, y), x), \text{sub}(x, \text{mult}(y, y))))$$

returns the positive square root of x if x is a squared integer. If x is not a squared integer, $f(x)$ is not defined.

2.1.3 *Kleene's normal form*(*)[7]

Let us stop along the way. Partial recursive functions are so diverse that at a glance, we cannot expect their general structure. However, in fact, here is an amazing theorem.

Theorem 2.1. (Kleene's normal form) *For any partial recursive function* $f : \mathbb{N}^n \to \mathbb{N}$, *there exist two primitive recursive functions* $g, p : \mathbb{N}^{n+1} \to \mathbb{N}$ *such that*

$$f(x_1, x_2, \ldots, x_n) = g(x_1, x_2, \ldots, x_n, \mu_y(p(x_1, x_2, \ldots, x_n, y))), \quad (x_1, x_2, \ldots, x_n) \in \mathbb{N}^n. \quad (2.1)$$

We here only give an idea of the proof by explaining an example. The point is that if a given program has more than one loop, which correspond to μ-operators, we can rearrange them into a single loop.

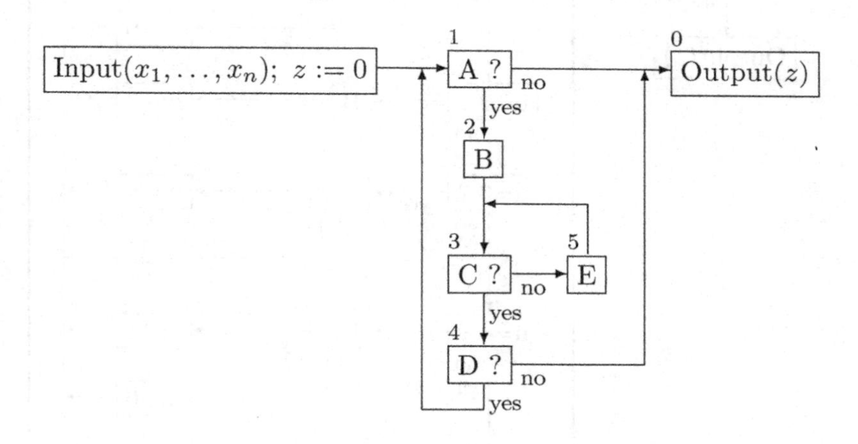

Fig. 2.2 Flow chart I [8]

Let Fig. 2.2 (Flow chart I) be a flow chart to compute the function f. [A ?] [C ?] [D ?] show conditions of branches, and [B] [E] are procedures without loops (i.e., calculation of primitive recursive functions), which set the value of z, respectively. This program includes the main loop [A ?] → [B] → [C ?] → [D ?] → [A ?], a nested loop [C ?] → [E] → [C ?], and an

[7] The subsections with (∗) can be skipped.

[8] Flow charts I, II are slight modifications of Fig. 3 (p.12), Fig. 4 (p.13) of [Takahashi (1991)], respectively.

escape branch from the main loop at D ?. Let us show that these loops can be rearranged into a single loop by introducing a new variable u. To do this, we put numbers 0 to 5 respectively at the top left of the boxes of all procedures A ? C ? D ? B E and the output procedure in order for the variable u to refer (Fig. 2.2).

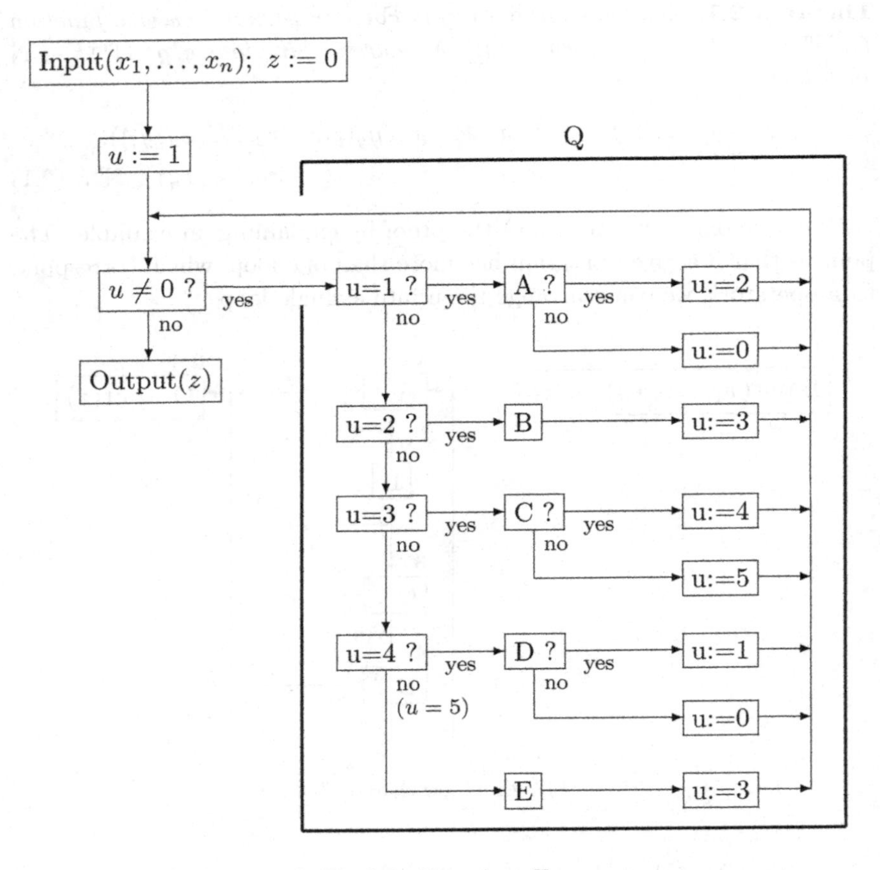

Fig. 2.3 Flow chart II

Fig. 2.3 (Flow chart II) shows the rearrangement of Flow chart I. It is easy to confirm that Flow chart II also computes the same function f. Let us show that f, which Flow chart II computes, can be expressed in the form (2.1). Let Q be a process consisting of all the procedures enclosed by the thick lines in Flow chart II. Define $g(x_1, \ldots, x_n, y)$ as the value of the output variable z that is produced after Q being executed y times under the input $(x_1, \ldots, x_n)$, and define $p(x_1, \ldots, x_n, y)$ as the value of u after Q being executed y times under the input $(x_1, \ldots, x_n)$. Then,we see (2.1) holds.

Kleene's normal form is introduced into the design of computer. Indeed, a software called compiler converts Flow chart I, which we can easily understand, to Flow chart II, which a computer can easily deal with. The number written at the top left of each box in Flow chart I corresponds to the address of memory where the procedure in the box is physically stored. The new variable u corresponds to the program counter of the central processing unit (CPU), which is used to indicate the address of memory that the CPU is looking at.

2.1.4 *Enumeration theorem*

The set of all partial recursive functions is a countable set, which we can actually enumerate.

Theorem 2.2. (Enumeration theorem) *There exists a partial recursive function*

$$\mathrm{univ}_n : \mathbb{N} \times \mathbb{N}^n \to \mathbb{N}$$

that has the following property: for each partial recursive function $f : \mathbb{N}^n \to \mathbb{N}$, there exists an $e_f \in \mathbb{N}$ such that

$$\mathrm{univ}_n(e_f, x_1, \ldots, x_n) = f(x_1, \ldots, x_n), \quad (x_1, \ldots, x_n) \in \mathbb{N}^n.$$

The function univ_n is called an *enumerating function* or a *universal function*, and e_f is called a *Gödel number* (or an *index*) of f. We here give only an idea of the proof of Theorem 2.2. First, write a computer program of a given partial recursive function f, and regard it as a $\{0,1\}$-sequence, or a non-negative integer. The Gödel number e_f is such an integer. The enumerating function $\mathrm{univ}_n(e, x_1, \ldots, x_n)$ checks if e is a Gödel number of some partial recursive function f of n variables. If it is not, $\mathrm{univ}_n(e, x_1, \ldots, x_n)$ is not defined. If it is, $\mathrm{univ}_n(e, x_1, \ldots, x_n)$ then reconstructs the partial recursive function f from $e = e_f$, and finally computes $f(x_1, \ldots, x_n)$.

A personal computer (PC), although it has only finite memory, can be regarded as a realization of an enumerating function. On the other hand, a partial recursive function is usually regarded as a single-purpose computer, such as a pocket calculator. If we install a program (= Gödel number) of pocket calculator into a PC, it quickly acts as a pocket calculator. In this sense, an enumerating function or a PC is 'universal'. Just like there is more than one type of PC, there is more than one enumerating function. Just like there is more than one program for a given function, there is more than one Gödel number for a given partial recursive function.

In order for the enumeration theorem to hold, the notion of 'partial function' is essential.

Theorem 2.3. *Any total function that is an extension of* univ_n *is not recursive.*

Proof. We show the theorem by contradiction.†[9] Suppose that there exists a total recursive function g that is an extension of univ_n. Then,

$$\varphi(z, x_2, \ldots, x_n) := g(z, z, x_2, \ldots, x_n) + 1 \tag{2.2}$$

is also a total recursive function. Therefore using a Gödel number e_φ of φ, we can write

$$\varphi(z, x_2, \ldots, x_n) = \mathrm{univ}_n(e_\varphi, z, x_2, \ldots, x_n).$$

Since φ is total, so is $\mathrm{univ}_n(e_\varphi, \cdot, \cdot, \ldots, \cdot)$. Since g is an extension of univ_n, we see

$$\varphi(z, x_2, \ldots, x_n) = g(e_\varphi, z, x_2, \ldots, x_n).$$

Putting $z = e_\varphi$ in the above equality, we get

$$\varphi(e_\varphi, x_2, \ldots, x_n) = g(e_\varphi, e_\varphi, x_2, \ldots, x_n).$$

On the other hand, (2.2) implies that

$$\varphi(e_\varphi, x_2, \ldots, x_n) = g(e_\varphi, e_\varphi, x_2, \ldots, x_n) + 1,$$

which is a contradiction. □

Let us introduce an important consequence of Theorem 2.3. Define a function $\mathrm{halt}_n : \mathbb{N} \times \mathbb{N}^n \to \{0, 1\}$ by

$$\mathrm{halt}_n(z, x_1, \ldots, x_n) := \begin{cases} 1 \ (\mathrm{univ}_n(z, x_1, \ldots, x_n) \text{ is defined}), \\ 0 \ (\mathrm{univ}_n(z, x_1, \ldots, x_n) \text{ is not defined}). \end{cases}$$

[9] In proving impossibilities seen in such theorems of computer science as this theorem or Theorem 2.7, self-referential versions of 'diagonal method' are used. The proof below is self-referential in that we substitute the Gödel number e_φ for the first argument of φ defined by (2.2).

halt_n is a total function. Let $f : \mathbb{N}^n \to \mathbb{N}$ be a partial recursive function and let e_f be its Gödel number. Then, $\mathrm{halt}_n(e_f, x_1, \ldots, x_n)$ judges whether $f(x_1, \ldots, x_n)$ is defined or not. Since 'not defined' means an infinite loop for actual computers, for any given f and $(x_1, \ldots, x_n)$, asking whether $f(x_1, \ldots, x_n)$ is defined or not is called the *halting problem.* Here is an important theorem.

Theorem 2.4. halt_n *is not a total recursive function.*

Proof. Consider a function $g : \mathbb{N} \times \mathbb{N}^n \to \mathbb{N}$ defined by

$$g(z, x_1, x_2, \ldots, x_n) := \begin{cases} \mathrm{univ}_n(z, x_1, x_2, \ldots, x_n) & (\mathrm{halt}_n(z, x_1, \ldots, x_n) = 1), \\ 0 & (\mathrm{halt}_n(z, x_1, \ldots, x_n) = 0). \end{cases}$$

If halt_n is a total recursive function, so is g, but this is impossible by Theorem 2.3 because g is an extension of univ_n. □

Theorem 2.4 implies that there is no program that computes halt_n. Namely, there is no program that judges whether an arbitrarily given program halts or fall into an infinite loop for an arbitrarily given input $(x_1, \ldots, x_n)$. In more familiar words, there is no program that judges whether an arbitrarily given program has a bug or not.

A little knowledge of number theory helps us feel Theorem 2.4 more familiar. Define a subset $B(x) \subset \mathbb{N}$, $x \in \mathbb{N}$, by

$$B(x) := \left\{ y \geqq x \;\middle|\; \begin{array}{l} y \text{ is a positive even integer that cannot be} \\ \text{expressed as a sum of two prime numbers} \end{array} \right\},$$

and a partial recursive function $f : \mathbb{N} \to \mathbb{N}$ by

$$f(x) := \begin{cases} \min B(x) & (B(x) \neq \emptyset), \\ \text{not defined} & (B(x) = \emptyset). \end{cases}$$

Then, if halt_1 were a total recursive function, there would exist a program that computes $\mathrm{halt}_1(e_f, 4)$, i.e., we would be able to know whether Goldbach's conjecture†[10] is true or not. Like this, the function halt_1 would solve many other unsolved problems in number theory. This is quite unlikely.

2.2 Kolmogorov complexity and random number

In Sec. 1.2, we mentioned that a long $\{0, 1\}$-sequence ω is called a random number if the shortest program q_ω to produce it is almost as long as ω itself. In this section, we exactly define random number in terms of recursive function.

[10] A conjecture that any even integer not less than 4 can be expressed as a sum of two prime numbers.

2.2.1 *Kolmogorov complexity*

Each finite $\{0,1\}$-sequence can be identified with a non-negative integer in the following way.

Definition 2.3. Let $\{0,1\}^* := \bigcup_{n\in\mathbb{N}}\{0,1\}^n$. Namely, $\{0,1\}^*$ is the set of all finite $\{0,1\}$-sequences. An element of $\{0,1\}^*$, i.e., a finite $\{0,1\}$-sequence, is called a *word*. In particular, the $\{0,1\}$-sequence of length 0 is called the *empty word*. The *canonical order* in $\{0,1\}^*$ is defined in the following way: for $x, y \in \{0,1\}^*$, if x is longer than y then define $x > y$, if x and y have a same length then define the order regarding them as binary integers. We identify $\{0,1\}^*$ with $\mathbb{N}$ by the canonical order. For example, the empty word$= 0$, $(0) = 1$, $(1) = 2$, $(0,0) = 3$, $(0,1) = 4$, $(1,0) = 5$, $(1,1) = 6$, $(0,0,0) = 7$, $\ldots$.

Definition 2.4. For each $q \in \{0,1\}^*$, let $L(q) \in \mathbb{N}$ denote the n such that $q \in \{0,1\}^n$, i.e., $L(q)$ is the length of q. For $q \in \mathbb{N}$, $L(q)$ means the length of the corresponding $\{0,1\}$-sequence to q in the canonical order. For example, $L(5) = L((1,0)) = 2$. In general, $L(q)$ is equal to the integer part of $\log_2(q+1)$. i.e., $L(q) = \lfloor \log_2(q+1) \rfloor$.

$L(q)$ is non-decreasing function. In particular,

$$L(q) \leqq \log_2 q + 1, \quad q \in \mathbb{N}_+.^{\dagger 11} \tag{2.3}$$

Now, we introduce a useful operation that makes a function of one variable from that of several variables. If $x_i \in \{0,1\}^{m_i}$, $i = 1, 2$, are

$$x_1 = (x_{11}, x_{12}, \ldots, x_{1m_1}), \quad x_2 = (x_{21}, x_{22}, \ldots, x_{2m_2}),$$

we define $\langle x_1, x_2 \rangle \in \{0,1\}^{2m_1+2+m_2}$ by

$$\langle x_1, x_2 \rangle := (x_{11}, x_{11}, x_{12}, x_{12}, \ldots, x_{1m_1}, x_{1m_1}, 0, 1, x_{21}, x_{22}, \ldots, x_{2m_2}). \tag{2.4}$$

By induction, we define

$$\langle x_1, x_2, \ldots, x_n \rangle := \langle x_1, \langle x_2, \ldots, x_n \rangle \rangle, \quad n = 3, 4, \ldots.$$

The inverse functions of $u = \langle x_1, \ldots, x_n \rangle$ are written as

$$(u)_i^n := x_i, \quad i = 1, 2, \ldots, n,$$

which are primitive recursive functions.

---

[11] $\mathbb{N}_+ := \{1, 2, \ldots\}$ is the set of all positive integers.

For example, for $(1) \in \{0,1\}^1$ and $(1,1,0,1,1) \in \{0,1\}^5$, we have

$$\langle (1), (1,1,0,1,1) \rangle = (1,1,0,1,1,1,0,1,1) =: u \in \{0,1\}^9.$$

Then, $(u)^2_1 = (1)$ and $(u)^2_2 = (1,1,0,1,1)$. At the same time, since $\langle (1),(1),(1) \rangle = u$, we have $(u)^3_1 = (u)^3_2 = (u)^3_3 = (1)$.

The function $\langle x_1, \ldots, x_n \rangle$ is a mathematical description of a way to hand more than one parameter to a function. For example, as is seen in a description like $f(2,3)$, a string of letters '2, 3' is put into $f(\cdot)$, each parameter being divided by ' , '. In computer, '2, 3' is coded as a certain word u. Since u has the information about ' , ', we can decode 2 and 3 from u. The delimiter ' , ' corresponds to '0, 1' in the definition (2.4) of $\langle x_1, x_2 \rangle$.

Definition 2.5. (Computational complexity depending on algorithm) Let $A : \{0,1\}^* \to \{0,1\}^*$ be a partial recursive function as a function $\mathbb{N} \to \mathbb{N}$. We call A an algorithm. The computational complexity of $x \in \{0,1\}^*$ under the algorithm A is defined by

$$K_A(x) := \min\{L(q) \,|\, q \in \{0,1\}^*,\ A(q) = x\}.$$

If there is no such q that $A(q) = x$, we set $K_A(x) := \infty$.

In Definition 2.5, Kolmogorov named A an algorithm, but today, it may be better to call it a programming language. The input q of A is then a program. Thus, $K_A(x)$ is interpreted as the length of the shortest program that computes x under the programming langauge A.

Since K_A depends on one particular algorithm A, it is not a universal index for complexity. Now, we introduce the following theorem.

Theorem 2.5. *There exists an algorithm $A_0 : \{0,1\}^* \to \{0,1\}^*$ such that for any algorithm $A : \{0,1\}^* \to \{0,1\}^*$, we can find such a constant $c_{A_0A} \in \mathbb{N}$ that*

$$K_{A_0}(x) \leqq K_A(x) + c_{A_0A}, \quad x \in \{0,1\}^*.$$

A_0 is called a universal algorithm.

Proof. Using an enumerating function univ_1, we define an algorithm A_0 by

$$A_0(z) := \text{univ}_1((z)^2_1, (z)^2_2), \qquad z \in \{0,1\}^*.$$

If z is not of the form $z = \langle e, q \rangle$, we do not define $A_0(z)$. If e_A is a Gödel number of A, we have $A_0(\langle e_A, q \rangle) = \text{univ}_1(e_A, q) = A(q)$. Take an arbitrary

$x \in \{0,1\}^*$. If there is no q such that $A(q) = x$, then $K_A(x) = \infty$ and the desired inequality holds. If there is such a q, let q_x be the shortest such one, i.e., $K_A(x) = L(q_x)$. Since $A_0(\langle e_A, q_x \rangle) = x$, we have

$$K_{A_0}(x) \leqq L(\langle e_A, q_x \rangle).$$

It follows from (2.4) that

$$L(\langle e_A, q_x \rangle) = L(q_x) + 2L(e_A) + 2 = K_A(x) + 2L(e_A) + 2.$$

Hence

$$K_{A_0}(x) \leqq K_A(x) + 2L(e_A) + 2, \quad x \in \{0,1\}^*.$$

Putting $c_{A_0 A} := 2L(e_A) + 2$, the theorem holds. □

When $K_{A_0}(x) \gg c_{A_0 A}$, if $K_{A_0}(x)$ is greater than $K_A(x)$, the difference is relatively small. Therefore for x such that $K_{A_0}(x) \gg c_{A_0 A}$, either $K_{A_0}(x)$ is less than $K_A(x)$ or $K_{A_0}(x)$ is relatively slightly greater than $K_A(x)$. In this sense, A_0 is also called an *asymptotically optimal algorithm*.

Let A_0 and A_0' be two universal algorithms.†[12] Then, putting $c := \max\{c_{A_0' A_0}, c_{A_0 A_0'}\}$, we have

$$\left| K_{A_0}(x) - K_{A_0'}(x) \right| < c, \quad x \in \{0,1\}^*. \tag{2.5}$$

This means that when $K_{A_0}(x)$ or $K_{A_0'}(x)$ is much greater than c, their difference can be ignored.

Definition 2.6. We fix a universal algorithm A_0, and define

$$K(x) := K_{A_0}(x), \quad x \in \{0,1\}^*.$$

We call $K(x)$ the *Kolmogorov complexity*†[13] of x.

Theorem 2.6. *(i) There exists a constant $c > 0$ such that*

$$K(x) \leqq n + c, \quad x \in \{0,1\}^n, \quad n \in \mathbb{N}_+.$$

In particular, $K : \{0,1\}^* \to \mathbb{N}$ is a total function.
(ii) If $n > c' > 0$, then we have

$$\#\{x \in \{0,1\}^n \mid K(x) \geqq n - c'\} > 2^n - 2^{n-c'}.$$

Proof. (i) For an algorithm $A(x) := \mathrm{proj}_1^1(x) = x$, we have $K_A(x) = n$ for $x \in \{0,1\}^n$. Consequently, Theorem 2.5 implies $K(x) \leqq n + c$ for some constant $c > 0$. (ii) The number of q such that $L(q) < n - c'$ is equal to $2^0 + 2^1 + \cdots + 2^{n-c'-1} = 2^{n-c'} - 1$, and hence the number of $x \in \{0,1\}^*$ such that $K(x) < n - c'$ is at most $2^{n-c'} - 1$. From this (ii) follows. □

[12] Since there is more than one enumerating function, there is more than one universal algorithm.
[13] It is also called the Kolmogorov-Chaitin complexity, algorithmic complexity, description complexity, ... etc.

2.2.2 *Random number*

For $n \gg 1$, we call $x \in \{0,1\}^n$ a *random number* if $K(x) \approx n$. Theorem 2.6 implies that when $n \gg 1$, nearly all $x \in \{0,1\}^n$ are random numbers.

Remark 2.1. Since there is more than one universal algorithm, and the differences among them include ambiguities (2.5), the definition of random number cannot help having some ambiguity.†[14]

Example 2.3. The world record of computation of π is 2,576,980,370,000 decimal digits or approximately 8,560,543,490,000 bits (as of August 2009). Since the program that produced the record is much shorter than this, the $\{0,1\}$-sequence of π in its binary expansion up to 8,560,543,490,000 digit is not a random number.

As is seen in Example 2.3, we know many $x \in \{0,1\}^*$ that are not random. However, we know no concrete example of random number. Indeed, the following theorem implies that there is no algorithm to judge whether a given $x \in \{0,1\}^n$, $n \gg 1$, is random or not.

Theorem 2.7. *The Kolmogorov complexity $K(x)$ is not a total recursive function.*

Proof. Let us identify $\{0,1\}^*$ with $\mathbb{N}$. We show the theorem by contradiction. Suppose that $K(x)$ is a total recursive function. Then, a function

$$\psi(x) := \min\{z \in \mathbb{N} \mid K(z) \geqq x\}, \quad x \in \mathbb{N},$$

is also a total recursive function. We see $x \leqq K(\psi(x))$ by definition. Considering ψ to be an algorithm, we have

$$K_\psi(\psi(x)) = \min\{L(q) \mid q \in \mathbb{N},\, \psi(q) = \psi(x)\},$$

and consequently, $K_\psi(\psi(x)) \leqq L(x)$. Therefore by Theorem 2.5, we know that there exists a constant $c > 0$ such that for any x,

$$x \leqq K(\psi(x)) \leqq L(x) + c. \tag{2.6}$$

However, (2.6) is impossible for $x \gg 1$ because $L(x) \leqq \log_2 x + 1$ ((2.3) and Proposition A.3 (ii) in case $a = 1$). Thus $K(x)$ is not a total recursive function. □

---

[14] We can also define random infinite $\{0,1\}$-sequences, i.e., *infinite random number*. In that case, there is no ambiguity.

$K(x)$ is a total function, i.e., it is defined for all $x \in \{0,1\}^*$, but there is no program that computes it. Thus 'definable' and 'computable' are different concepts.

Theorem 2.7 is deeply related to Theorem 2.4. Consider a function complexity(x) defined below. In its definition, A_0 is the universal algorithm appeared in the proof of Theorem 2.5, which is assumed to be used for the definition of $K(x)$.

complexity(x)

(1) $l := 0$.
(2) Let q be the first word of $\{0,1\}^l$.
(3) If $A_0(q) = x$, then output l and halt.
(4) If q is the last word of $\{0,1\}^l$, increase l by 1, and go to (2).
(5) Assign q the next word of it, and go to (3).

Starting from the shortest program, the function complexity(x) executes every program q to check whether it computes x or not, and if it does, complexity(x) halts with output $K(x)$. However this program does not necessarily halt. Indeed, for some x, it must fall into an infinite loop at step (3) before $K(x)$ is computed, which cannot be avoided in advance because of Theorem 2.4.†15

Suppose, for example, that a large photo image of beautiful scenery is stored in a computer as a long word x. Since it is far from a random image, x is by no means a random number, i.e., $K(x) \ll L(x)$. This means that there is a $q_x \in \{0,1\}^*$ such that $A_0(q_x) = x$ and $K(x) = L(q_x)$. Then, storing q_x instead of x considerably saves the computer memory. This is the principle of *data compression*. The raw data x is compressed into q_x, and A_0 develops q_x to x. Unfortunately, we cannot compute q_x from x because $K(x)$ is not computable. In practice, some alternative methods are used for data compression.†16

2.2.3 *Application: Distribution of prime numbers*$^{(*)}$

Let us stop along the way, again. The Kolmogorov complexity has many applications not only in probability theory but also in other fields of mathematics ([Li and Vitányi (2008)]). We here present one of their applications

15 It is known that incomputability of the halting problem and that of the Kolmogorov complexity are equivalent.

16 One of such methods will be used in the proof of Theorem 3.1.

in number theory. First, we look at *Euclid's theorem*:

Theorem 2.8. *There are infinitely many prime numbers.*

Proof. We show the theorem by contradiction. Suppose that there are only finitely many prime numbers, say $p_1, p_2, \ldots, p_k$. Then, define a primitive recursive function $A : \mathbb{N} \to \mathbb{N}$ by

$$A(\langle e_1, e_2, \ldots, e_k \rangle) := p_1^{e_1} p_2^{e_2} \cdots p_k^{e_k}.$$

For an arbitrary $m \in \mathbb{N}_+$, there exist $e_1(m), e_2(m), \ldots, e_k(m) \in \mathbb{N}$ such that

$$A(\langle e_1(m), e_2(m), \ldots, e_k(m) \rangle) = m.$$

Consequently,

$$\begin{aligned} K_A(m) &\leqq L(\langle e_1(m), \ldots, e_k(m) \rangle) \\ &= 2L(e_1(m)) + 2 + \cdots + 2L(e_{k-1}(m)) + 2 + L(e_k(m)). \end{aligned}$$

For each i, we have $e_i(m) \leqq \log_{p_i} m \leqq \log_2 m$, and hence $L(e_i(m)) \leqq \log_2(\log_2 m + 1)$. This shows that

$$K_A(m) \leqq (2k-1)\log_2(\log_2 m + 1) + 2(k-1).$$

Therefore there exists a $c \in \mathbb{N}_+$ such that

$$K(m) \leqq (2k-1)\log_2(\log_2 m + 1) + 2(k-1) + c, \quad m \in \mathbb{N}_+.$$

If $m \gg 1$ is a random number, $K(m) \approx \log_2 m$. Thus the above inequality does not hold for large random numbers m, which is a contradiction. □

Of course, this proof is much more difficult than the well-known proof of Euclid's. However, polishing it, we can unexpectedly get a deep interesting knowledge about the distribution of prime numbers.

Theorem 2.9. ([Hardy and Wright (1979)], Theorem 8) *Let p_n be the n-th smallest prime number. Then, as $n \to \infty$, we have*

$$p_n \sim n \log n. \tag{2.7}$$

Here '$\sim$' indicates that the ratio of both sides converges to 1.

Theorem 2.9 is a paraphrase of the well-known *prime number theorem*: as $n \to \infty$,

$$\#\{2 \leqq p \leqq n \,|\, p \text{ is a prime number}\} \sim \frac{n}{\log n}.$$

In what follows, we approach Theorem 2.9 by using the Kolmogorov complexity, although the discussion below is somewhat lacking in rigor.

Let p_n be a divisor of $m \in \mathbb{N}_+$. Then, we can compute m from the two integers n and m/p_n.[†17] Therefore, $c, c', c'', c''' \in \mathbb{N}_+$ being constants that are independent of n and m,

$$\begin{aligned} K(m) &\leqq L\left(\left\langle n, \frac{m}{p_n}\right\rangle\right) + c \\ &= 2L(n) + 2 + L\left(\frac{m}{p_n}\right) + c \\ &\leqq 2\log_2 n + \log_2 m - \log_2 p_n + c'. \end{aligned} \tag{2.8}$$

If m is a random number, we have $K(m) \geqq \log_2 m - c''$, and hence

$$\log_2 p_n \leqq 2\log_2 n + c''',$$

namely,

$$p_n \leqq 2^{c'''} n^2.$$

This inequality is much weaker than (2.7). Let us improve it by describing the information about n and m/p_n as shortly as possible.

For $x = (x_1, \ldots, x_k) \in \{0,1\}^*$ and $y = (y_1, \ldots, y_l) \in \{0,1\}^*$, we define their concatenation xy by

$$xy := (x_1, \ldots, x_k, y_1, \ldots, y_l).$$

For more than two words, we define their concatenation similarly. It is impossible to decode x and y from xy uniquely because xy includes no delimiter. To decode them, we consider $\langle L(x), xy\rangle$. This time, since the length of x is known, we can decode x and y from xy uniquely. The length of the word $\langle L(x), xy\rangle$ is

$$L(\langle L(x), xy\rangle) = 2L(L(x)) + 2 + L(x) + L(y),$$

which is much shorter than $L(\langle x, y\rangle)$ for $x \gg 1$. This is again more compressed into

$$\langle L(L(x)), L(x)xy\rangle,$$

whose length is estimated as

$$L(\langle L(L(x)), L(x)xy\rangle) \approx 2\log_2\log_2\log_2 x + 2 + \log_2\log_2 x + \log_2 x + \log_2 y. \tag{2.9}$$

[17]The function $n \mapsto p_n$ is a primitive recursive function.

Substituting (2.9) for the inequality (2.8), we get the following estimate: $c, c', c'', c''' \in \mathbb{N}_+$ being constants that are independent of n and m,

$$K(m) \leqq 2\log_2\log_2\log_2 n + 2 + \log_2\log_2 n + \log_2 n + \log_2 m - \log_2 p_n + c.$$

If m is a random number, we have $K(m) \geqq \log_2 m - c'$, and hence

$$\log_2 p_n \leqq 2\log_2\log_2\log_2 n + \log_2\log_2 n + \log_2 n + c'',$$

namely,

$$p_n \leqq c''' n \log_2 n\,(\log_2\log_2 n)^2. \tag{2.10}$$

Although (2.10) shows only an upper bound with an unknown constant $c''' \in \mathbb{N}_+$, its increasing rate is very close to that of (2.7). It is amazing that (2.10) is obtained only by the definition of random number and a rough estimate of the Kolmogorov complexity.

Chapter 3

Limit theorem

It is extremely important to study and understand limit theorems because they are concrete means and expression forms to explore randomness, i.e., properties of random numbers. To explore randomness, the main purpose of the study of limit theorems is to discover as many non-trivial examples of events as possible whose probabilities are very close to 1.

Many practical applications of probability theory make use of limit theorems. For limit theorems that specify events whose probabilities are close to 1 enable us to predict future with very high probability even in random phenomena. Therefore limit theorems are the central theme in both theoretical and practical aspects.

In this chapter, we study two of the most important limit theorems for coin tosses, i.e., Bernoulli's theorem and de Moivre–Laplace's theorem. Furthermore, we learn that these two limit theorems extend to the law of large numbers and the central limit theorem, respectively, for arbitrary sequences of independent identically distributed random variables.

Even in the forefront of probability theory, these two theorems are extending in many ways and in diverse situations.

3.1 Bernoulli's theorem

When we toss a coin, it comes up Heads with probability $1/2$ and Tails with probability $1/2$, but this does not mean that the number of Heads and the number of Tails in 100 coin tosses are always equal to 50 and 50. The following is a real record of Heads($= 1$) and Tails($= 0$) of the author's trial of 100 coin tosses.

1110110101 1011101101 0100000011 0110101001 0101000100
0101111101 1010000000 1010100011 0100011001 1101111101

The number of 1's in the above sequence, i.e., the number of Heads in the 100 coin tosses, is 51. Another trial will bring a different sequence, but the number of Heads seldom deviates from 50 so much. In fact, when $n \gg 1$, if $\omega \in \{0,1\}^n$ is a random number, the relative frequency of 1's is approximately 1/2. This follows from Theorem 3.1 as we will explain on p.45.

This section exclusively deals with n coin tosses; recall that P_n is the uniform probability measure on $\{0,1\}^n$, and $\xi_i : \{0,1\}^n \to \mathbb{R}$, $i = 1, \ldots, n$, are the coordinate functions defined by (1.12). Then, the random variables $\xi_1, \ldots, \xi_n$ defined on the probability space $(\{0,1\}^n, \mathfrak{P}(\{0,1\}^n), P_n)$ represent n coin tosses.

Theorem 3.1. *There exists a constant $c \in \mathbb{N}_+$, such that for any $n \in \mathbb{N}_+$ and any $\omega \in \{0,1\}^n$, putting $p := \sum_{i=1}^n \xi_i(\omega)/n$, we have*

$$K(\omega) \leqq nH(p) + 4\log_2 n + c.$$

Here $H(p)$ is the (binary) entropy function*:*

$$H(p) := \begin{cases} -p\log_2 p - (1-p)\log_2(1-p) & (0 < p < 1), \\ 0 & (p = 0, 1). \end{cases} \tag{3.1}$$

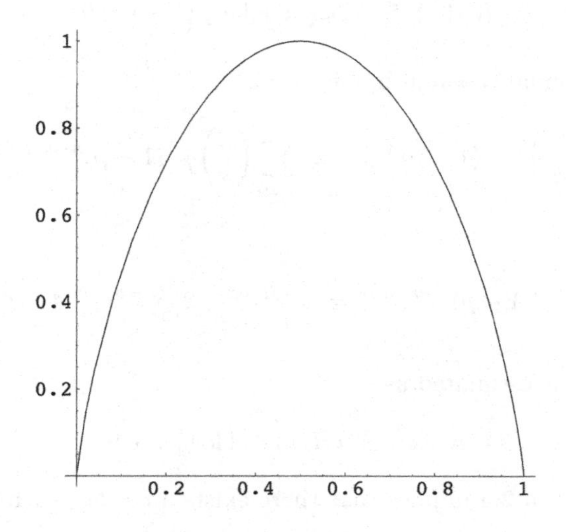

Fig. 3.1 The graph of $H(p)$

Proof. The number of $y \in \{0,1\}^n$ that satisfy $\sum_{i=1}^n \xi_i(y) = np$ is

$$\binom{n}{np} = \frac{n!}{(n-np)!(np)!}.$$

Suppose that among those y, the given ω is the n_1-th smallest in the canonical order. Let $A : \{0,1\}^* \to \{0,1\}^*$ be an algorithm that computes from $\langle n, m, l \rangle$ the l-th smallest word in the canonical order among the words $y \in \{0,1\}^n$ that satisfy $\sum_{i=1}^n \xi_i(y) = m$. Then, we have

$$A(\langle n, np, n_1 \rangle) = \omega.$$

Consequently,

$$\begin{aligned} K_A(\omega) &= L(\langle n, np, n_1 \rangle) \\ &= 2L(n) + 2 + 2L(np) + 2 + L(n_1) \\ &\leqq 4L(n) + L(n_1) + 4 \\ &\leqq 4L(n) + L\left(\binom{n}{np}\right) + 4. \end{aligned}$$

Using (2.3), we obtain

$$K_A(\omega) \leqq 4\log_2 n + \log_2 \binom{n}{np} + 9.$$

Now, the binomial theorem implies that

$$\binom{n}{np} p^{np}(1-p)^{n-np} \leqq \sum_{k=0}^{n} \binom{n}{k} p^k (1-p)^{n-k} = 1,$$

and hence

$$\binom{n}{np} \leqq p^{-np}(1-p)^{-(n-np)} = 2^{-np\log_2 p}\, 2^{-(n-np)\log_2(1-p)} = 2^{nH(p)}. \tag{3.2}$$

Then, $K_A(\omega)$ is estimated as

$$K_A(\omega) \leqq nH(p) + 4\log_2 n + 9. \tag{3.3}$$

Finally, Theorem 2.5 implies that there exists a $c \in \mathbb{N}_+$ such that

$$K(\omega) \leqq nH(p) + 4\log_2 n + c.$$

□

The entropy function $H(p)$ has the following properties (Fig. 3.1).

(i) $0 \leqq H(p) \leqq 1$. $H(p)$ takes its maximum value 1 at $p = 1/2$.

(ii) For $0 < \varepsilon < 1/2$, we have $0 < H\left(\frac{1}{2} + \varepsilon\right) = H\left(\frac{1}{2} - \varepsilon\right) < 1$, the common value, which is written as $H\left(\frac{1}{2} \pm \varepsilon\right)$ below, decreases as ε increases.

In the proof of Theorem 3.1, if $L(\omega) = n \gg 1$ and $p \neq 1/2$, we showed that the word ω can be compressed into the shorter word $\langle n, np, n_1\rangle$. This means that ω is not a random number. Taking the contrapositive, we can say that if $L(\omega) \gg 1$ and ω is a random number, then $\sum_{i=1}^{n} \xi_i(\omega)/n$ should be close to 1/2.

Theorem 3.1 implies *Bernoulli's theorem* (Theorem 3.2).†1

Theorem 3.2. *For any $\varepsilon > 0$, it holds that*

$$\lim_{n\to\infty} P_n\left(\left|\frac{\xi_1 + \cdots + \xi_n}{n} - \frac{1}{2}\right| > \varepsilon\right) = 0. \tag{3.4}$$

Remark 3.1. If $0 < \varepsilon < \varepsilon'$ and if (3.4) holds for ε, it also holds for ε'. Namely, the smaller ε is, the stronger the assertion (3.4) becomes. Therefore "For any $\varepsilon > 0$" in Theorem 3.2 actually means "No matter how small $\varepsilon > 0$ is". Such descriptions will often be seen below.

Proof of Theorem 3.2. Theorem 3.1 and the property (ii) of $H(p)$ imply that

$$\frac{\xi_1(\omega) + \cdots + \xi_n(\omega)}{n} > \frac{1}{2} + \varepsilon \Longrightarrow K(\omega) < nH\left(\frac{1}{2} \pm \varepsilon\right) + 4\log_2 n + c,$$

$$\frac{\xi_1(\omega) + \cdots + \xi_n(\omega)}{n} < \frac{1}{2} - \varepsilon \Longrightarrow K(\omega) < nH\left(\frac{1}{2} \pm \varepsilon\right) + 4\log_2 n + c.$$

Consequently, we have

$$\begin{aligned}
&\left\{\omega \in \{0,1\}^n \;\middle|\; \left|\frac{\xi_1(\omega) + \cdots + \xi_n(\omega)}{n} - \frac{1}{2}\right| > \varepsilon\right\} \\
&\quad = \left\{\omega \in \{0,1\}^n \;\middle|\; \frac{\xi_1(\omega) + \cdots + \xi_n(\omega)}{n} > \frac{1}{2} + \varepsilon\right\} \\
&\qquad \cup \left\{\omega \in \{0,1\}^n \;\middle|\; \frac{\xi_1(\omega) + \cdots + \xi_n(\omega)}{n} < \frac{1}{2} - \varepsilon\right\} \\
&\quad \subset \left\{\omega \in \{0,1\}^n \;\middle|\; K(\omega) < nH\left(\frac{1}{2} \pm \varepsilon\right) + 4\log_2 n + c\right\} \\
&\quad \subset \left\{\omega \in \{0,1\}^n \;\middle|\; K(\omega) < \left\lfloor nH\left(\frac{1}{2} \pm \varepsilon\right) + 4\log_2 n + c + 1\right\rfloor\right\}.
\end{aligned} \tag{3.5}$$

[1]Bernoulli's theorem includes the case of *unfair* coin tosses (Example 3.4). Theorem 3.2 deals with its special case.

By Theorem 2.6 (ii), we have

$$\begin{aligned}
&\#\left\{ \omega \in \{0,1\}^n \,\middle|\, \left| \frac{\xi_1(\omega)+\cdots+\xi_n(\omega)}{n} - \frac{1}{2} \right| > \varepsilon \right\} \\
&\leqq \#\left\{ \omega \in \{0,1\}^n \,\middle|\, K(\omega) < \left\lfloor nH\left(\frac{1}{2} \pm \varepsilon\right) + 4\log_2 n + c + 1 \right\rfloor \right\} \\
&\leqq 2^{\lfloor nH(\frac{1}{2}\pm\varepsilon)+4\log_2 n+c+1\rfloor} \leqq n^4 2^{c+1} 2^{nH(\frac{1}{2}\pm\varepsilon)}.
\end{aligned}$$

Dividing the both sides by $\#\{0,1\}^n = 2^n$, it holds that

$$P_n\left(\left| \frac{\xi_1+\cdots+\xi_n}{n} - \frac{1}{2} \right| > \varepsilon \right) \leqq n^4 2^{c+1} 2^{-n\left(1-H(\frac{1}{2}\pm\varepsilon)\right)}. \tag{3.6}$$

Since $1 - H\left(\frac{1}{2} \pm \varepsilon\right) > 0$, the right-hand side of (3.6) converges to 0 as $n \to \infty$ (see Proposition A.3 (i)). □

Let $c' \in \mathbb{N}_+$ be a constant, and let $A_n := \{\omega \in \{0,1\}^n \mid K(\omega) > n - c'\}$. Taking complements of the both sides of (3.5), if $n \gg 1$, we have

$$\begin{aligned}
A_n &\subset \left\{ \omega \in \{0,1\}^n \,\middle|\, K(\omega) \geqq \left\lfloor nH\left(\frac{1}{2} \pm \varepsilon\right) + 4\log_2 n + c + 1 \right\rfloor \right\} \\
&\subset \left\{ \omega \in \{0,1\}^n \,\middle|\, \left| \frac{\xi_1(\omega)+\cdots+\xi_n(\omega)}{n} - \frac{1}{2} \right| \leqq \varepsilon \right\} =: B_n.
\end{aligned}$$

A_n is considered to be the set of random numbers of length n, which is included by an event B_n of probability close to 1 that Theorem 3.2 specifies. Thus, in this case, Fig. 1.3 is like Fig. 3.2.

Here we derived a limit theorem (Theorem 3.2) directly from a property of random numbers (Theorem 3.1). In general, such derivation is difficult, and it is more usual to derive properties of randomness indirectly from limit theorems (Sec. 1.3.1).

Remark 3.2. Since Theorem 2.6 (ii) is also valid for the computational complexity K_A depending on A, according to (3.3), the constant c of (3.6) can be taken as 9. Then, putting $c = 9$, $\varepsilon = 1/2000$ and $n = 10^8$, (3.6) is now

$$P_{10^8}\left(\left| \frac{\xi_1+\cdots+\xi_{10^8}}{10^8} - \frac{1}{2} \right| > \frac{1}{2000} \right) \leqq 1.9502 \times 10^{13}. \tag{3.7}$$

Since the left-hand side is less than 1, this is unfortunately a meaningless inequality.

In the following sections, the inequality (3.7) will be much improved by the strong power of calculus. See (3.15), (3.22) and Exampe 3.10.

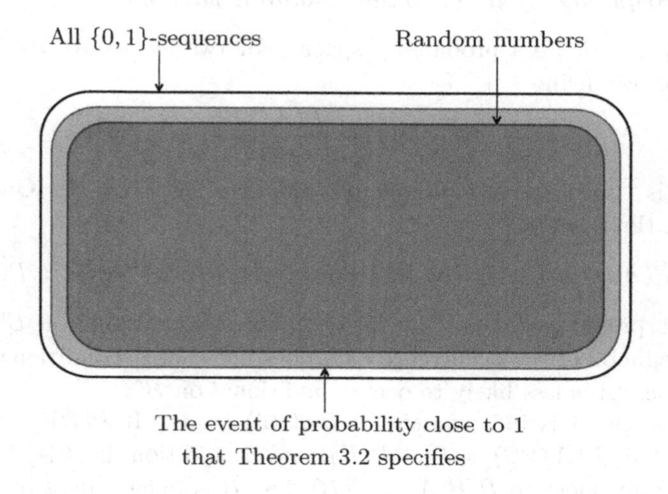

Fig. 3.2 Theorem 3.2 and random numbers (Conceptual figure)

3.2 Law of large numbers

The assertion of Bernoulli's theorem is valid not only for coin tosses, but also for more general sequences of random variables. Such extensions of Bernoulli's theorem are called the law of large numbers. In this book, we present the most basic case—the law of large numbers for sequences of independent identically distributed random variables.

Although it is called 'law', it is of course a mathematical theorem.†2 The law of large numbers, which entered the stage when probability was not recognized in mathematics, was considered to be a natural law like the law of inertia.

It took more than 200 years since Bernoulli's age for probability to be recognized in mathematics. As the 6-th of his 23 problems, Hilbert proposed a problem "axiomatic treatment of probability with limit theorems for foundation of statistical physics" at the Paris conference of the International Congress of Mathematicians in 1900. Probability has been recognized widely in mathematics since Borel presented the normal number theorem (1.16) in 1909. Finally, the axiomatization was done by [Kolomogorov (1933)].

[2]Poisson named it the law of large numbers.

3.2.1 *Sequence of independent random variables*

Let $(\Omega, \mathfrak{P}(\Omega), P)$ be a probability space. For two events $A, B \in \mathfrak{P}(\Omega)$, if $P(B) > 0$, we define

$$P(A|B) := \frac{P(A \cap B)}{P(B)}.$$

$P(A|B)$ is called the *conditional probability of A given B*. One of the following three holds:

$$\text{(i) } P(A|B) > P(A), \quad \text{(ii) } P(A|B) = P(A), \quad \text{(iii) } P(A|B) < P(A).$$

(i) is interpreted as "A is more likely to occur conditional on B", (ii) as "The likelihood of the occurrence of A does not change conditional on B", and (iii) as "A is less likely to occur conditional on B".

Let us say A is independent of B if (ii) holds. If $P(B) > 0$, (ii) is equivalent to $P(A \cap B) = P(A)P(B)$, and in addition, if $P(A) > 0$, then it is also equivalent to $P(B|A) = P(B)$, i.e., B is independent of A. Thus independence of two events is symmetric. Making the symmetry clear and including the cases $P(A) = 0$ or $P(B) = 0$, we say A and B are independent if

$$P(A \cap B) = P(A)P(B).$$

In general, independence of n events is defined as follows.

Definition 3.1. We say events $A_1, \ldots, A_n$ are *independent* if for any subset $I \subset \{1, \ldots, n\}$, $I \neq \emptyset$,

$$P\left(\bigcap_{i \in I} A_i\right) = \prod_{i \in I} P(A_i)$$

holds. Here $\bigcap_{i \in I} A_i$ denotes the intersection of A_i for all $i \in I$, and $\prod_{i \in I} P(A_i)$ denotes the product of $P(A_i)$ for all $i \in I$ (Sec. A.1.2).

Remark 3.3. On the probability space $(\{0,1\}^2, \mathfrak{P}(\{0,1\}^2), P_2)$, consider the following three events

$$A := \{(0,0),(0,1)\}, \quad B := \{(0,1),(1,1)\}, \quad C := \{(0,0),(1,1)\}.$$

Then, since

$$P_2(A \cap B) = P_2(A)P_2(B) = \frac{1}{4},$$

$$P_2(B \cap C) = P_2(B)P_2(C) = \frac{1}{4},$$

$$P_2(C \cap A) = P_2(C)P_2(A) = \frac{1}{4},$$

any two out of the three are independent. However, since

$$P_2(A \cap B \cap C) = P_2(\emptyset) = 0, \quad P_2(A)P_2(B)P_2(C) = \frac{1}{8},$$

A, B, C are not independent.

Definition 3.2. (i) If random variables X and Y (not necessarily defined on a same probability space) have a same distribution, we say they are *identically distributed.* If random variables $X_1, \ldots, X_n$ and $Y_1, \ldots, Y_n$ (not necessarily defined on a same probability space) have a same joint distribution, we say they are *identically distributed.*
(ii) We say random variables $X_1, \ldots, X_n$ are *independent* if for any $c_i < d_i$, $i = 1, \ldots, n$, the events

$$\{ c_i < X_i \leqq d_i \}, \quad i = 1, \ldots, n$$

are independent.

Events $A_1, \ldots, A_n \in \mathfrak{P}(\Omega)$ are independent if and only if their indicator functions $\mathbf{1}_{A_1}, \ldots, \mathbf{1}_{A_n}$ (Sec. A.1.1) are independent as random variables.

Proposition 3.1. *Let $X_1, \ldots, X_n$ be random variables, and let $\{ a_{ij} \mid j = 1, \ldots, s_i \}$ be the range of each X_i, $i = 1, \ldots, n$. The following two are equivalent.*
(i) $X_1, \ldots, X_n$ are independent.
(ii) It holds that

$$P(X_i = a_{ij_i},\ i = 1, \ldots, n) = \prod_{i=1}^{n} P(X_i = a_{ij_i})$$

for any $j_i = 1, \ldots, s_i$, $i = 1, \ldots, n$.

Proof. (i) $\Longrightarrow$ (ii): If we take c_i and d_i so that $\{c_i < X_i \leqq d_i\} = \{X_i = a_{ij_i}\}$ holds, (i) implies that $\{X_i = a_{ij_i}\}$, $i = 1, \ldots, n$, are independent, i.e., (ii) follows.
(ii) $\Longrightarrow$ (i): We have

$$P(c_1 < X_1 \leqq d_1) = \sum_{j_1\, ;\, c_1 < a_{1j_1} \leqq d_1} P(X_1 = a_{1j_1}),$$

where $\sum_{j_1\, ;\, c_1 < a_{1j_1} \leqq d_1}$ denotes the sum over all j_1 such that $c_1 < a_{1j_1} \leqq d_1$

(Sec. A.1.2). Similarly,

$$
\begin{aligned}
&P(c_i < X_i \leqq d_i,\ i = 1, \ldots, n) \\
&= \sum_{j_1\,;\,c_1 < a_{1j_1} \leqq d_1} \cdots \sum_{j_n\,;\,c_n < a_{nj_n} \leqq d_n} P(\,X_i = a_{ij_i},\ i = 1, \ldots, n\,) \\
&= \sum_{j_i\,;\,c_1 < a_{1j_1} \leqq d_1} \cdots \sum_{j_n\,;\,c_n < a_{nj_n} \leqq d_n} \prod_{i=1}^{n} P(\,X_i = a_{ij_i}\,).
\end{aligned}
$$

Resolving it into factors (Example A.2), we obtain

$$
\begin{aligned}
&= \prod_{i=1}^{n} \sum_{j_i\,;\,c_i < a_{ij_i} \leqq d_i} P(\,X_i = a_{ij_i}\,) \\
&= \prod_{i=1}^{n} P(\,c_i < X_i \leqq d_i\,).
\end{aligned}
$$

Take any subset $I \subset \{1, 2, \ldots, n\}$, $I \neq \emptyset$. For each $k \notin I$, if we put $c_k := \min\{a_{kj} \mid j = 1, 2, \ldots, s_k\} - 1$, and $d_k := \max\{a_{kj} \mid j = 1, 2, \ldots, s_k\}$, we have

$$
P(c_k < X_k \leqq d_k) = 1, \quad k \notin I,
$$

and hence

$$
\begin{aligned}
P(c_i < X_i \leqq d_i,\ i \in I) &= P(c_i < X_i \leqq d_i,\ i = 1, 2, \ldots, n) \\
&= \prod_{i=1}^{n} P(\,c_i < X_i \leqq d_i\,) \\
&= \prod_{i \in I} P(\,c_i < X_i \leqq d_i\,).
\end{aligned}
$$

□

Example 3.1. The coordinate functions $\xi_1, \ldots, \xi_n$ defined on the probability space $(\{0,1\}^n, \mathfrak{P}(\{0,1\}^n), P_n)$ are independent identically distributed (abbreviated as *i.i.d.*) random variables (Example 1.4).

Proposition 3.1 (ii) shows that if $X_1, \ldots, X_n$ are independent, their joint distribution can be constructed of their marginal distributions. Suppose that for each $i = 1, \ldots, n$, a random variable $X_i : \Omega_i \to \mathbb{R}$ is defined on a probability space $(\Omega_i, \mathfrak{P}(\Omega_i), \mu_i)$. Then, on a suitable probability space $(\hat{\Omega}, \mathfrak{P}(\hat{\Omega}), \hat{\mu})$, we can construct independent random variables $\hat{X}_1, \ldots, \hat{X}_n$

so that X_i and $\hat{X}_i$ are identically distributed for each $i = 1,\ldots,n$. For example,

$$\hat{\Omega} := \Omega_1 \times \cdots \times \Omega_n,$$
$$\hat{\mu}(\{\omega\}) := \prod_{i=1}^{n} \mu_i(\{\omega_i\}), \quad \omega = (\omega_1,\ldots,\omega_n) \in \hat{\Omega},$$
$$\hat{X}_i(\omega) := X_i(\omega_i), \quad \omega = (\omega_1,\ldots,\omega_n) \in \hat{\Omega}, \quad i = 1,\ldots,n.$$

Here the probability measure $\hat{\mu}$ is written as

$$\hat{\mu} = \mu_1 \otimes \cdots \otimes \mu_n,$$

and is called the *product probability measure* of $\mu_1,\ldots,\mu_n$.

Example 3.2. Let us construct a mathematical model of general coin tosses that assumes the probabilities of Heads$(=1)$ and Tails$(=0)$ are $0 < p < 1$ and $1-p$, respectively. If $p \neq 1/2$, it is a model of *unfair coin tosses*.

For each $1 \leqq i \leqq n$, define a probability space $(\Omega_i, \mathfrak{P}(\Omega_i), \mu_i)$ by

$$\Omega_i := \{0,1\}, \quad \mu_i(\{1\}) := p, \quad \mu_i(\{0\}) = 1-p.$$

Then, we construct a probability space $(\hat{\Omega}, \mathfrak{P}(\hat{\Omega}), P_n^{(p)})$, where

$$\hat{\Omega} := \Omega_1 \times \cdots \times \Omega_n = \{0,1\}^n,$$
$$P_n^{(p)} := \mu_1 \otimes \cdots \otimes \mu_n.$$

On this probability space, the coordinate functions $\{\xi_i\}_{i=1}^n$ defined by (1.12) represent n coin tosses, which are unfair if $p \neq 1/2$. Namely, $\{\xi_i\}_{i=1}^n$ are i.i.d. with

$$P_n^{(p)}(\xi_i = 1) = p, \quad P_n^{(p)}(\xi_i = 0) = 1-p, \quad i = 1,\ldots,n.$$

The product probability measure $P_n^{(p)}$ is given by

$$P_n^{(p)}(\{\omega\}) := p^{\omega_1+\cdots+\omega_n}(1-p)^{n-(\omega_1+\cdots+\omega_n)}, \quad \omega = (\omega_1,\ldots,\omega_n) \in \{0,1\}^n.$$

In particular, $P_n^{(1/2)} = P_n$.

Definition 3.3. Let $(\Omega, \mathfrak{P}(\Omega), P)$ be a probability space, and $X : \Omega \to \mathbb{R}$ be a random variable defined on it. We define the *mean* (or the *expectation*) $\mathbf{E}[X]$ of X and the *variance* $\mathbf{V}[X]$ of X by

$$\mathbf{E}[X] := \sum_{\omega \in \Omega} X(\omega) P(\{\omega\}),$$
$$\mathbf{V}[X] := \mathbf{E}\left[(X - \mathbf{E}[X])^2\right] = \sum_{\omega \in \Omega} (X(\omega) - \mathbf{E}[X])^2 P(\{\omega\}).$$

Proposition 3.2. *Let X be a random variable, and let $\{a_1, a_2, \ldots, a_s\}$ be the range of X. Then, we have*

$$\mathbf{E}[X] = \sum_{i=1}^{s} a_i P(X = a_i),$$

$$\mathbf{V}[X] = \sum_{i=1}^{s} (a_i - \mathbf{E}[X])^2 P(X = a_i).$$

In particular, for an event $A \in \mathfrak{P}(\Omega)$, we have $\mathbf{E}[\,\mathbf{1}_A\,] = P(A)$.

Proof.

$$\begin{aligned}
\mathbf{E}[X] &= \sum_{\omega \in \Omega} X(\omega) P(\{\omega\}) \\
&= \sum_{i=1}^{s} \sum_{\omega \in \Omega\,;\, X(\omega) = a_i} X(\omega) P(\{\omega\}) \quad \text{(Sec. A.1.2)} \\
&= \sum_{i=1}^{s} \sum_{\omega \in \Omega\,;\, X(\omega) = a_i} a_i P(\{\omega\}) \\
&= \sum_{i=1}^{s} a_i \sum_{\omega \in \Omega\,;\, X(\omega) = a_i} P(\{\omega\}) \\
&= \sum_{i=1}^{s} a_i P(X = a_i).
\end{aligned}$$

The proof is similar for $\mathbf{V}[X]$. □

By Proposition 3.2, the mean and the variance of a random variable are determined by its distribution. Therefore if X and Y are identically distributed, their means and variances coincide, respectively.

Proposition 3.3. *Let $(\Omega, \mathfrak{P}(\Omega), P)$ be a probability space. For any random variables $X_1, \ldots, X_n : \Omega \to \mathbb{R}$ and any constants $c_1, \ldots, c_n \in \mathbb{R}$, it holds that*

$$\mathbf{E}[c_1 X_1 + \cdots + c_n X_n] = c_1 \mathbf{E}[X_1] + \cdots + c_n \mathbf{E}[X_n]. \tag{3.8}$$

Proof.

$$\begin{aligned}
\mathbf{E}[c_1 X_1 + \cdots + c_n X_n] &= \sum_{\omega \in \Omega} (c_1 X_1(\omega) + \cdots + c_n X_n(\omega))\, P(\{\omega\}) \\
&= c_1 \sum_{\omega \in \Omega} X_1(\omega) P(\{\omega\}) + \cdots + c_n \sum_{\omega \in \Omega} X_n(\omega) P(\{\omega\}) \\
&= c_1 \mathbf{E}[X_1] + \cdots + c_n \mathbf{E}[X_n].
\end{aligned}$$

□

Proposition 3.4. *The variance of a random variable X can be calculated as follows.*

$$\mathbf{V}[X] = \mathbf{E}\left[X^2\right] - \mathbf{E}[X]^2.$$

Proof. By Proposition 3.3, we have

$$\begin{aligned}\mathbf{V}[X] &= \mathbf{E}\left[X^2 - 2X\mathbf{E}[X] + \mathbf{E}[X]^2\right] \\ &= \mathbf{E}\left[X^2\right] - \mathbf{E}\left[2X\mathbf{E}[X]\right] + \mathbf{E}\left[\mathbf{E}[X]^2\right] \\ &= \mathbf{E}\left[X^2\right] - 2\mathbf{E}[X]\mathbf{E}[X] + \mathbf{E}[X]^2 \\ &= \mathbf{E}\left[X^2\right] - \mathbf{E}[X]^2.\end{aligned}$$

□

Proposition 3.5. *Let $(\Omega, \mathfrak{P}(\Omega), P)$ be a probability space. For any independent random variables $X_1, \ldots, X_n : \Omega \to \mathbb{R}$ and any constants $c_1, \ldots, c_n \in \mathbb{R}$, it holds that*

$$\mathbf{E}[X_1 \times \cdots \times X_n] = \mathbf{E}[X_1] \times \cdots \times \mathbf{E}[X_n], \tag{3.9}$$

$$\mathbf{V}[c_1 X_1 + \cdots + c_n X_n] = c_1^2 \mathbf{V}[X_1] + \cdots + c_n^2 \mathbf{V}[X_n]. \tag{3.10}$$

Proof. For $i = 1, \ldots, n$, let $\{a_{ij} \mid j = 1, \ldots, s_i\}$ be the range of X_i. By Proposition 3.2, we have

$$\begin{aligned}&\mathbf{E}[X_1 \times \cdots \times X_n] \\ &= \sum_{j_1=1}^{s_1} \cdots \sum_{j_n=1}^{s_n} a_{1j_1} \times \cdots \times a_{nj_n} P(X_1 = a_{1j_1}, \ldots, X_n = a_{nj_n}) \\ &= \sum_{j_1=1}^{s_1} \cdots \sum_{j_n=1}^{s_n} a_{1j_1} \times \cdots \times a_{nj_n} P(X_1 = a_{1j_1}) \times \cdots \times P(X_n = a_{nj_n}) \\ &= \sum_{j_1=1}^{s_1} a_{1j_1} P(X_1 = a_{1j_1}) \times \cdots \times \sum_{j_n=1}^{s_n} a_{nj_n} P(X_n = a_{nj_n}) \\ &= \mathbf{E}[X_1] \times \cdots \times \mathbf{E}[X_n],\end{aligned}$$

thus (3.9) is proved. For the variance, we have

$$\begin{aligned}
&\mathbf{V}[c_1X_1+\cdots+c_nX_n]\\
&=\mathbf{E}\left[(c_1X_1+\cdots+c_nX_n-\mathbf{E}[c_1X_1+\cdots+c_nX_n])^2\right]\\
&=\mathbf{E}\left[(c_1(X_1-\mathbf{E}[X_1])+\cdots+c_n(X_n-\mathbf{E}[X_n]))^2\right]\\
&=\sum_{i=1}^n c_i^2\mathbf{E}\left[(X_i-\mathbf{E}[X_i])^2\right]+\sum_{i\neq j}c_ic_j\mathbf{E}\left[(X_i-\mathbf{E}[X_i])(X_j-\mathbf{E}[X_j])\right]\\
&=\sum_{i=1}^n c_i^2\mathbf{V}[X_i]+\sum_{i\neq j}c_ic_j\mathbf{E}\left[(X_i-\mathbf{E}[X_i])(X_j-\mathbf{E}[X_j])\right].
\end{aligned}$$

Since X_i and X_j are independent if $i \neq j$, so are $X_i-\mathbf{E}[X_i]$ and $X_j-\mathbf{E}[X_j]$. Hence

$$\begin{aligned}
\mathbf{E}[(X_i-\mathbf{E}[X_i])(X_j-\mathbf{E}[X_j])] &= \mathbf{E}[(X_i-\mathbf{E}[X_i])]\mathbf{E}[(X_j-\mathbf{E}[X_j])]\\
&= 0\times 0 = 0,
\end{aligned}$$

from which (3.10) follows. □

Remark 3.4. In order for (3.10) to hold, the assumption "$X_1,\ldots,X_n$ are independent" is somewhat too much, and an assumption "X_i and X_j are independent if $i \neq j$ (*pairwise independent*)" suffices. For example, in Remark 3.3, the indicator functions $\mathbf{1}_A, \mathbf{1}_B$ and $\mathbf{1}_C$ of the events A, B, and C are not independent, but they are pairwise independent.

3.2.2 *Chebyshev's inequality*

In general, for any non-negative random variable $X \geqq 0$, the following inequality holds.

$$P(X \geqq a) \leqq \frac{\mathbf{E}[X]}{a}, \quad a>0. \tag{3.11}$$

This is called *Markov's inequality*. Indeed, if $\{a_1,a_2,\ldots,a_s\}$, $a_i \geqq 0$, is the range of X, Proposition 3.2 implies that

$$\begin{aligned}
\mathbf{E}[X] &= \sum_{i=1}^s a_iP(X=a_i)\\
&= \sum_{i\,;\,0\leqq a_i<a} a_iP(X=a_i)+\sum_{i\,;\,a_i\geqq a} a_iP(X=a_i)\\
&\geqq \sum_{i\,;\,a_i\geqq a} a_iP(X=a_i)\\
&\geqq a\sum_{i\,;\,a_i\geqq a} P(X=a_i) = aP(X\geqq a).
\end{aligned}$$

Dividing the both sides by a, we obtain (3.11).

Lemma 3.1. *The following inequality holds.*

$$P(\,|X-\mathbf{E}[X]|\geqq\varepsilon\,)\leqq\frac{\mathbf{V}[X]}{\varepsilon^2},\qquad\varepsilon>0.$$

This is called Chebyshev's inequality.

Proof. Applying Markov's inequality to $Y:=(X-\mathbf{E}[X])^2$, we obtain

$$P(\,|X-\mathbf{E}[X]|\geqq\varepsilon\,)=P(Y\geqq\varepsilon^2)\leqq\frac{\mathbf{E}[Y]}{\varepsilon^2}=\frac{\mathbf{V}[X]}{\varepsilon^2}.$$

□

Suppose that for each $n\in\mathbb{N}_+$, a probability space $(\Omega_n,\mathfrak{P}(\Omega_n),\mu_n)$, and i.i.d. random variables $X_{n,1},X_{n,2},\ldots,X_{n,n}$ are given. Suppose further that their common mean and variance are

$$\mathbf{E}[X_{n,k}]=m_n,\quad\mathbf{V}[X_{n,k}]=\sigma_n^2\leqq\sigma^2,\quad k=1,\ldots,n,$$

where σ^2 is a constant that does not depend on n. Then, Proposition 3.3 and Proposition 3.5 imply that

$$\mathbf{E}\left[\frac{X_{n,1}+\cdots+X_{n,n}}{n}\right]=m_n,\tag{3.12}$$

$$\mathbf{V}\left[\frac{X_{n,1}+\cdots+X_{n,n}}{n}\right]=\frac{\sigma_n^2}{n}\leqq\frac{\sigma^2}{n}.\tag{3.13}$$

Applying Chebyshev's inequality (Lemma 3.1), we have

$$\mu_n\left(\left|\frac{X_{n,1}+\cdots+X_{n,n}}{n}-m_n\right|\geqq\varepsilon\right)\leqq\frac{\sigma^2}{n\varepsilon^2},\qquad\varepsilon>0.$$

From this, the next theorem immediately follows.

Theorem 3.3. *For any* $\varepsilon>0$, *we have*

$$\lim_{n\to\infty}\mu_n\left(\left|\frac{X_{n,1}+\cdots+X_{n,n}}{n}-m_n\right|\geqq\varepsilon\right)=0.$$

Theorems, like Theorem 3.3, that assert "the distribution of the arithmetic mean of n identically distributed random variables concentrates to their common mean as $n\to\infty$" are called the *law of large numbers.*[†3]

[3]More exactly, Theorem 3.3 is called the *weak* law of large numbers to distinguish it from the *strong* law of large numbers. The prototype of the latter is Borel's normal number theorem (1.16).

Example 3.3. Applying Theorem 3.3 to the case where

$$(\Omega_n, \mathfrak{P}(\Omega_n), \mu_n) = (\{0,1\}^n, \mathfrak{P}(\{0,1\}^n), P_n)$$

and

$$X_{n,k}(\omega) := \xi_k(\omega) = \omega_k, \quad \omega = (\omega_1, \ldots, \omega_n) \in \{0,1\}^n, \qquad k = 1, \ldots, n,$$

we obtain Theorem 3.2. In this case, Chebyshev's inequality shows

$$P_n\left(\left| \frac{\xi_1 + \cdots + \xi_n}{n} - \frac{1}{2} \right| \geqq \varepsilon \right) \leqq \frac{1}{4n\varepsilon^2}. \tag{3.14}$$

In particular, putting $\varepsilon = 1/2000$ and $n = 10^8$, we obtain

$$P_{10^8}\left(\left| \frac{\xi_1 + \cdots + \xi_{10^8}}{10^8} - \frac{1}{2} \right| \geqq \frac{1}{2000} \right) \leqq \frac{1}{100}. \tag{3.15}$$

Compared with (3.7) being meaningless, the inequality (3.15) has a practical meaning.

Example 3.4. Let $0 < p < 1$. In Example 3.2, we introduced the probability space $(\{0,1\}^n, \mathfrak{P}(\{0,1\}^n), P_n^{(p)})$, and the coordinate functions $\{\xi_k\}_{k=1}^n$ as n coin tosses, which are unfair if $p \neq 1/2$. Since we have

$$\mathbf{E}[\xi_k] = p$$

and Proposition 3.4 implies

$$\mathbf{V}[\xi_k] = \mathbf{E}\left[\xi_k^2\right] - \mathbf{E}[\xi_k]^2 = p - p^2 = p(1-p),$$

it follows from Theorem 3.3 that

$$\lim_{n\to\infty} P_n^{(p)}\left(\left| \frac{\xi_1 + \cdots + \xi_n}{n} - p \right| \geqq \varepsilon \right) = 0, \quad \varepsilon > 0.$$

As we can imagine from Theorem 2.6 (ii), Theorem 3.1 and Example 3.4, it is known that for $n \gg 1$, the probability that $K(\omega) \approx nH(p)$, $\omega \in \{0,1\}^n$, is close to 1 under the probability measure $P_n^{(p)}$:

$$\lim_{n\to\infty} P_n^{(p)}\left(\left\{ \omega \in \{0,1\}^n \;\middle|\; \left| \frac{K(\omega)}{n} - H(p) \right| \geqq \varepsilon \right\} \right) = 0, \quad \varepsilon > 0.$$

If $H(p)$ is not extremely small, for $n \gg 1$, $nH(p)$ is a huge number so that even a computer cannot produce any $\omega \in \{0,1\}^n$ with $K(\omega) \approx nH(p)$. Strictly speaking, such ω is not a random number, but we may say it is random in usual sense. Consequently, a long $\{0,1\}$-sequence generated by unfair coin tosses may well look random with high probability.

Example 3.5. On the probability space $(\{0,1\}^{2n}, \mathfrak{P}(\{0,1\}^{2n}), P_{2n})$, let us consider the following random variables.

$$X_k(\omega) := \xi_{2k-1}(\omega)\xi_{2k}(\omega), \quad \omega \in \{0,1\}^{2n}, \quad k = 1, \ldots, n.$$

Here $\{\xi_k\}_{k=1}^{2n}$ are the coordinate functions. Then, $\{X_k\}_{k=1}^{n}$ are independent, and we have

$$P_{2n}(X_k = 1) = \frac{1}{4}, \quad P_{2n}(X_k = 0) = \frac{3}{4}, \quad k = 1, \ldots, n.$$

Namely, $\{X_k\}_{k=1}^{n}$ are n unfair coin tosses with $p = 1/4$. Therefore by Example 3.4, we see

$$\lim_{n\to\infty} P_{2n}\left(\left|\frac{X_1 + \cdots + X_n}{n} - \frac{1}{4}\right| \geqq \varepsilon\right) = 0, \quad \varepsilon > 0.$$

This means that in fair $(p = 1/2)$ coin tosses, the relative frequency of the occurrences of 'Heads, Heads $(= 11)$' approaches $1/4$ as $n \to \infty$ with high probability.

According to Kolmogorov's axioms of probability theory, if a triplet $(\Omega, \mathfrak{P}(\Omega), P)$ satisfies the required mathematical conditions of probability space, we call P a probability measure, and $P(A)$ the probability that A occurs, no matter whether it is related to a random phenomenon or not. Before his axiomatization, the question "What is probability?" had been so serious. Among answers, many people supported the following 'definition'. Let A be an event that is obtained as an outcome of a certain trial. Repeating the trials many times, if the relative frequency of the occurrences of A is considered to converge to a constant p, we 'define' the probability of A as this value p. This is called the *empirical probability.* The law of large numbers gave a mathematical ground to the idea of the empirical probability.

3.2.3 *Cramér–Chernoff's inequality*

Bernoulli's theorem (Theorem 3.2) was proved by the inequalities (3.6) or (3.14). The former is very loose for small n but converges to 0 exponentially fast as $n \to \infty$, while the latter gives a better estimate than the former for small n but it converges to 0 only as fast as $1/n$ as $n \to \infty$, and hence it gives a worse estimate than the former for $n \gg 1$. In this subsection, we prove an inequality (3.21), which improves both (3.6) and (3.14).

Applying Markov's inequality (3.11) to the exponential function e^{tX} of a random variable X, we have for any $t > 0$,

$$P(X \geqq x) = P\left(e^{tX} \geqq e^{tx}\right) \leqq \frac{\mathbf{E}\left[e^{tX}\right]}{e^{tx}} = M_X(t)\, e^{-tx}, \quad x \in \mathbb{R}. \quad (3.16)$$

Here

$$M_X(t) := \mathbf{E}\left[e^{tX}\right], \quad t \in \mathbb{R},$$

is called the *moment generating function* of X. Since $P(X \geqq x) \leqq M_X(t)\, e^{-tx}$ clearly holds for $t = 0$, (3.16) will be the most improved if we select $t \geqq 0$ so that the right-hand side takes its minimum value. Namely, the left-hand side of (3.16) is bounded from above by the minimum value of the function $M_X(t)\, e^{-tx}$ over all $t \geqq 0$:

$$P(X \geqq x) \leqq \min_{t \geqq 0} M_X(t)\, e^{-tx}, \quad x \in \mathbb{R}. \quad (3.17)$$

We call this inequality *Cramér–Chernoff's inequality*.

Moment generating function originated with [Lapalce (1812)]. The k-th derivative of $M_X(t)$ is

$$M_X^{(k)}(t) := \frac{d^k}{dt^k} \mathbf{E}\left[e^{tX}\right] = \mathbf{E}\left[\frac{d^k}{dt^k} e^{tX}\right] = \mathbf{E}\left[X^k e^{tX}\right].$$

Putting $t = 0$, we obtain

$$M_X^{(k)}(0) = \mathbf{E}\left[X^k\right], \quad k \in \mathbb{N}_+. \quad (3.18)$$

$\mathbf{E}\left[X^k\right]$ is called the k-th *moment* of X. Thus, as its name indicates, $M_X(t)$ generates every moment of X.

Let us apply Cramér–Chernoff's inequality to the sum of i.i.d. random variables.

Theorem 3.4. *For i.i.d. random variables $X_1, \ldots, X_n$ defined on a probability space $(\Omega, \mathfrak{P}(\Omega), P)$, it holds that*

$$P(X_1 + \cdots + X_n \geqq nx) \leqq \exp\left(-nI(x)\right), \quad x \in \mathbb{R}. \quad (3.19)$$

Here

$$I(x) := \max_{t \geqq 0} \left(tx - \log M_{X_1}(t)\right), ^{\dagger 4} \quad (3.20)$$

where $\max_{t \geqq 0} u(t)$ is the maximum value of $u(t)$ over all $t \geqq 0$.

[4] Since the function $I(x)$ in (3.20) characterizes the rate of decay of (3.19), it is called the *rate function.* Deriving a rate function from a moment generating function by the formula (3.20) is called the *Legendre transformation,* which plays important roles in convex analysis and analytical mechanics.

Proof. By Cramér–Chernoff's inequality,

$$\begin{aligned}
P(X_1 + \cdots + X_n \geqq nx) &\leqq \min_{t \geqq 0} \mathbf{E}\left[e^{t(X_1+\cdots+X_n)}\right] e^{-tnx} \\
&= \min_{t \geqq 0} \mathbf{E}\left[e^{tX_1} \times \cdots \times e^{tX_n}\right] e^{-tnx} \\
&= \min_{t \geqq 0} \mathbf{E}\left[e^{tX_1}\right] \times \cdots \times \mathbf{E}\left[e^{tX_n}\right] e^{-tnx} \\
&= \min_{t \geqq 0} M_{X_1}(t)^n e^{-tnx} \\
&= \min_{t \geqq 0} \exp\left(-n\left(tx - \log M_{X_1}(t)\right)\right) \\
&= \exp\left(-n \max_{t \geqq 0}\left(tx - \log M_{X_1}(t)\right)\right).
\end{aligned}$$

□

Let us apply Theorem 3.4 to the sum of n coin tosses $\xi_1 + \cdots + \xi_n$. Since

$$M_{\xi_1}(t) = e^{t \times 0} \times P_n(\xi_1 = 0) + e^{t \times 1} \times P_n(\xi_1 = 1) = \frac{1}{2}(1 + e^t),$$

to calculate $I(x)$, fixing x, we have to find the maximum value of

$$g(t) := tx - \log \frac{1}{2}(1 + e^t), \quad t \geqq 0.$$

Here let us assume

$$\frac{1}{2} < x < 1.$$

We first note that $g(0) = 0$ and $\lim_{t \to \infty} g(t) = -\infty$. By differentiating $g(t)$, we determine its increase and decrease. Solving

$$g'(t) = x - \frac{e^t}{1 + e^t} = 0,$$

we get

$$e^t = \frac{x}{1 - x}, \quad t = \log\left(\frac{x}{1 - x}\right) > 0.$$

Since $g''(s) = -e^s/(1 + e^s)^2 < 0$, g is a concave function, and hence the above solution t gives the maximum value of g. Thus

$$\begin{aligned}
I(x) &= g\left(\log \frac{x}{1 - x}\right) \\
&= x \log x + (1 - x)\log(1 - x) + \log 2.
\end{aligned}$$

Here the entropy function $H(p)$ appears, i.e.,

$$I(x) = -H(x)\log 2 + \log 2.$$

Therefore (3.19) is now

$$P_n(\xi_1 + \cdots + \xi_n \geqq nx) \leqq 2^{-n(1-H(x))}.$$

In particular, putting $x = \frac{1}{2} + \varepsilon$, $0 < \varepsilon < \frac{1}{2}$, we see

$$P_n\left(\frac{\xi_1 + \cdots + \xi_n}{n} \geqq \frac{1}{2} + \varepsilon\right) \leqq 2^{-n\left(1-H\left(\frac{1}{2}\pm\varepsilon\right)\right)}.$$

Exchanging 1 and 0 in the coin tosses, we have

$$P_n\left(\frac{\xi_1 + \cdots + \xi_n}{n} \leqq \frac{1}{2} - \varepsilon\right) = P_n\left(\frac{\xi_1 + \cdots + \xi_n}{n} \geqq \frac{1}{2} + \varepsilon\right).$$

Consequently, we finally obtain

$$P_n\left(\left|\frac{\xi_1 + \cdots + \xi_n}{n} - \frac{1}{2}\right| \geqq \varepsilon\right) \leqq 2 \cdot 2^{-n\left(1-H\left(\frac{1}{2}\pm\varepsilon\right)\right)}. \qquad (3.21)$$

This is an improvement of both (3.6) and (3.14).

Example 3.6. Putting $\varepsilon = 1/2000$ and $n = 10^8$ in (3.21), we obtain

$$P_{10^8}\left(\left|\frac{\xi_1 + \cdots + \xi_{10^8}}{10^8} - \frac{1}{2}\right| \geqq \frac{1}{2000}\right) \leqq 3.85747 \times 10^{-22}. \qquad (3.22)$$

This inequality is much more precise than Chebyshev's inequality (3.15).

Remark 3.5. Note that $\min_{t \geqq 0} u(t)$ or $\max_{t \geqq 0} u(t)$ does not always exist. For example, the former does not exist for $u(t) = 1/(1+t)$. In this book, we deal with only cases where they exist.

3.3 De Moivre–Laplace's theorem

The inequality (3.21) derived from Cramér–Chernoff's inequality can be more improved. The ultimate form (Example 3.10) is shown by de Moivre–Laplace's theorem.

3.3.1 *Binomial distribution*

Theorem 3.5. (Binomial distribution) *We use the setup of general coin tosses introduced in Example 3.2. Then, we have*

$$P_n^{(p)}(\xi_1 + \cdots + \xi_n = k) = \binom{n}{k} p^k (1-p)^{n-k}, \quad k = 0, 1, \ldots, n, \qquad (3.23)$$

in particular, if $p = 1/2$,

$$P_n(\xi_1 + \cdots + \xi_n = k) = \binom{n}{k} 2^{-n}, \quad k = 0, 1, \ldots, n.$$

Proof. We prove the theorem using the moment generating function of $\xi_1 + \cdots + \xi_n$. For each $t \in \mathbb{R}$,

$$\begin{aligned}
\mathbf{E}\left[e^{t(\xi_1+\cdots+\xi_n)}\right] &= \mathbf{E}\left[e^{t\xi_1} \times \cdots \times e^{t\xi_n}\right] \\
&= \mathbf{E}\left[e^{t\xi_1}\right] \times \cdots \times \mathbf{E}\left[e^{t\,\xi_n}\right] \\
&= \mathbf{E}\left[e^{t\xi_1}\right]^n \\
&= \left(e^0 P_n^{(p)}(\xi_1 = 0) + e^t P_n^{(p)}(\xi_1 = 1)\right)^n \\
&= \left((1-p) + e^t p\right)^n \\
&= \sum_{k=0}^{n} \binom{n}{k} \left(e^t p\right)^k (1-p)^{n-k} \\
&= \sum_{k=0}^{n} e^{tk} \binom{n}{k} p^k (1-p)^{n-k}.
\end{aligned}$$

On the other hand,

$$\mathbf{E}\left[e^{t(\xi_1+\cdots+\xi_n)}\right] = \sum_{k=0}^{n} e^{tk}\, P_n^{(p)}\left(\xi_1 + \cdots + \xi_n = k\right).$$

Comparing the coefficients of e^{tk} in the two expressions, we obtain (3.23). □

Figure 3.3 shows the distributions of $\xi_1 + \cdots + \xi_n$ ($n = 30, 100$) for $p = 1/2$. The rectangles (columns) composing the histogram are

$$\left[k - \frac{1}{2}, k + \frac{1}{2}\right) \times \left[0, \binom{n}{k} 2^{-n}\right], \quad k = 0, 1, \ldots, n,$$

(Example A.1). The area of the k-th rectangle is $P_n\left(\xi_1 + \cdots + \xi_n = k\right) = \binom{n}{k} 2^{-n}$, and the sum of the areas of all rectangles is 1.

3.3.2 *Heuristic observation*

Let $p = 1/2$. Then, we have $\mathbf{E}[\xi_i] = 1/2$ and $\mathbf{V}[\xi_i] = 1/4$, and hence (3.12) and (3.13) imply that

$$\mathbf{E}\left[\xi_1 + \cdots + \xi_n\right] = \frac{n}{2}, \qquad \mathbf{V}\left[\xi_1 + \cdots + \xi_n\right] = \frac{n}{4}.$$

From these, it follows that

$$Y_n(\alpha) := \frac{\left(\xi_1 - \frac{1}{2}\right) + \cdots + \left(\xi_n - \frac{1}{2}\right)}{n^\alpha}, \quad \alpha > 0$$

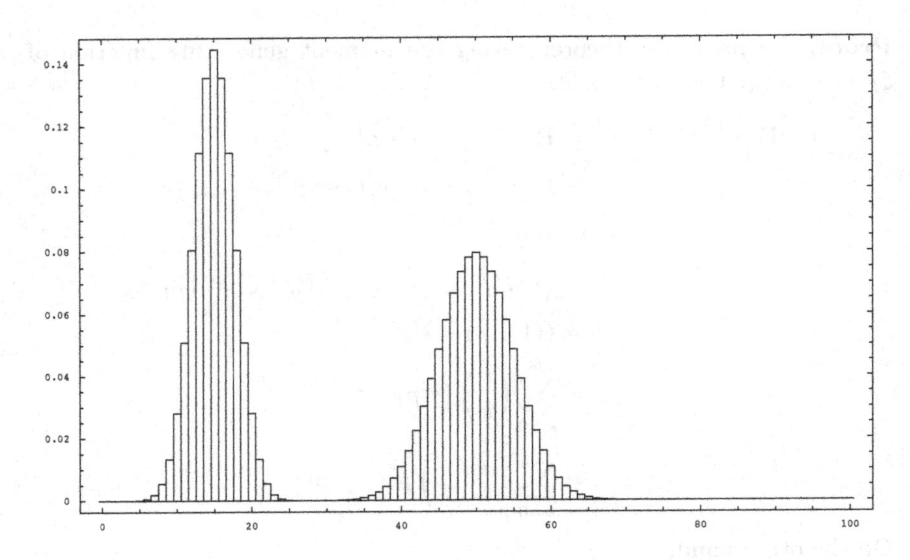

Fig. 3.3 The histogram of binomial distribution (Left: $n = 30$, Right: $n = 100$)

satisfies $\mathbf{E}[Y_n(\alpha)] = 0$ and $\mathbf{V}[Y_n(\alpha)] = 1/(4n^{2\alpha-1})$. In particular, if $\alpha > 1/2$, we have $\lim_{n\to\infty} \mathbf{V}[Y_n(\alpha)] = 0$, and hence, by Chebyshev's inequality,

$$P_n\left(|Y_n(\alpha)| \geqq \varepsilon\right) \leqq \frac{1}{4n^{2\alpha-1}\varepsilon^2} \to 0, \quad n \to \infty.$$

When $\alpha = 1$, this is Bernoulli's theorem.

Now, what happens if $\alpha = 1/2$? Modifying $Y_n(1/2)$ a little, we define

$$Z_n := \frac{\left(\xi_1 - \frac{1}{2}\right) + \cdots + \left(\xi_n - \frac{1}{2}\right)}{\frac{1}{2}\sqrt{n}}. \tag{3.24}$$

Then, we see $\mathbf{E}[Z_n] = 0$ and $\mathbf{V}[Z_n] = 1$. In general, for a random variable X, the mean and the variance of

$$\frac{X - \mathbf{E}[X]}{\sqrt{\mathbf{V}[X]}}$$

are 0 and 1, respectively. This is called the *standardization* of X.

The histogram of the distribution of Z_n seems to be convergent to a smooth function as $n \to \infty$ (Fig. 3.4). Sacrificing rigor a little, let us find the limit function heuristically.

For $x \in \mathbb{R}$ satisfying

$$x = \frac{r - \frac{n}{2}}{\frac{1}{2}\sqrt{n}} \tag{3.25}$$

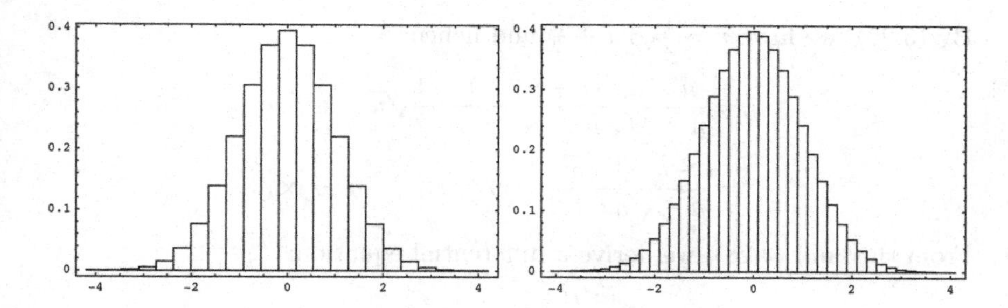

Fig. 3.4 The histogram of the distribution of Z_n (Left: $n = 30$, Right: $n = 100$)

with some $r = 0, 1, \ldots, n$, the probability that $Z_n = x$ is given by

$$P_n(Z_n = x) = \binom{n}{r} 2^{-n}.$$

The area of the column of the histogram standing on x is the probability $P_n(Z_n = x)$, and its width is $1/\left(\frac{1}{2}\sqrt{n}\right)$, and hence its height $f_n(x)$ is given by

$$f_n(x) = P_n(Z_n = x) \div \frac{1}{\frac{1}{2}\sqrt{n}} = \binom{n}{r} 2^{-n} \cdot \frac{1}{2}\sqrt{n}. \tag{3.26}$$

Now, consider the following ratio:

$$\frac{f_n(x) - f_n\left(x - \frac{1}{\frac{1}{2}\sqrt{n}}\right)}{\frac{1}{\frac{1}{2}\sqrt{n}}} \div f_n(x). \tag{3.27}$$

If $f_n(x)$ converges to a smooth function $f(x)$ as $n \to \infty$, we may well expect that

$$(3.27) \longrightarrow \frac{f'(x)}{f(x)}, \qquad n \to \infty. \tag{3.28}$$

(3.27) can be calculated, by using (3.26), as

$$\begin{aligned}
\left(1 - \frac{f_n\left(x - \frac{1}{\frac{1}{2}\sqrt{n}}\right)}{f_n(x)}\right) \frac{1}{2}\sqrt{n} &= \left(1 - \frac{\binom{n}{r-1} 2^{-n}}{\binom{n}{r} 2^{-n}}\right) \frac{1}{2}\sqrt{n} \\
&= \left(1 - \frac{r}{n-r+1}\right) \frac{1}{2}\sqrt{n} \\
&= \frac{n-2r+1}{n-r+1} \cdot \frac{1}{2}\sqrt{n}.
\end{aligned}$$

By (3.25), we have $r = \frac{1}{2}\sqrt{n}x + \frac{n}{2}$ and hence

$$= \frac{n - (\sqrt{n}x + n) + 1}{n - \frac{1}{2}(\sqrt{n}x + n) + 1} \cdot \frac{1}{2}\sqrt{n}$$
$$= \frac{-nx + \sqrt{n}}{n - \sqrt{n}x + 2} \longrightarrow -x, \qquad n \to \infty.$$

From this and (3.28), we derive a differential equation:

$$\frac{f'(x)}{f(x)} = -x.$$

Integrating the both sides, we obtain

$$\log f(x) = -\frac{x^2}{2} + C \qquad (C \text{ is an integral constant}).$$

Namely,

$$f(x) = \exp\left(-\frac{x^2}{2} + C\right) = e^C \exp\left(-\frac{x^2}{2}\right).$$

By (3.25) and (3.26), the constant e^C should be

$$e^C = f(0) = \lim_{n\to\infty} f_{2n}(0) = \lim_{n\to\infty} \binom{2n}{n} 2^{-2n} \cdot \frac{1}{2}\sqrt{2n}.$$

By Wallis' formula (Corollary 3.1), we see $e^C = 1/\sqrt{2\pi}$, and hence we obtain

$$f(x) = \frac{1}{\sqrt{2\pi}} \exp\left(-\frac{x^2}{2}\right). \tag{3.29}$$

This is the density function of the standard normal distribution (or the standard Gaussian distribution, Fig. 3.5).

In the above reasoning, (3.28) has no logical basis. Nevertheless, in fact, *de Moivre–Laplace's theorem* (Theorem 3.6) holds.†5

Theorem 3.6. *Let Z_n be the standardized sum of n coin tosses defined by (3.24). Then, for any real numbers $A < B$, it holds that*

$$\lim_{n\to\infty} P_n(A \leqq Z_n \leqq B) = \int_A^B \frac{1}{\sqrt{2\pi}} \exp\left(-\frac{x^2}{2}\right) dx.$$

We will prove Theorem 3.6 in Sec. 3.3.4.

[5] De Moivre–Laplace's theorem includes the case of unfair coin tosses (Example 3.2). Theorem 3.6 is a special case of it.

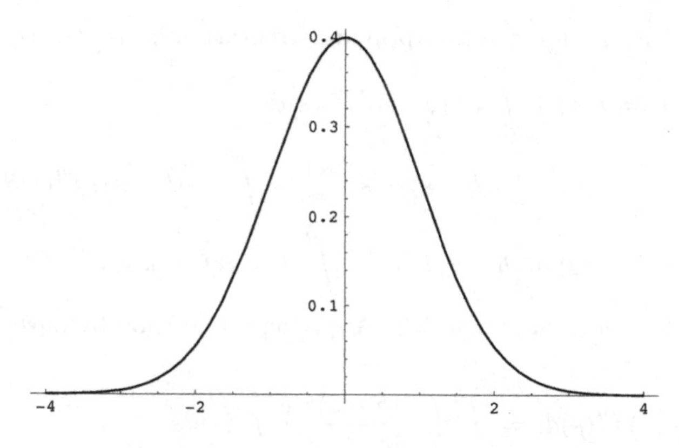

Fig. 3.5 The graph of $\frac{1}{\sqrt{2\pi}}\exp\left(-\frac{x^2}{2}\right)$

3.3.3 *Taylor's formula and Stirling's formula*

In order to prove de Moivre–Laplace's theorem, we introduce two basic formulas of calculus.

Theorem 3.7. (Taylor's formula) *If a function f is n times continuously differentiable in an interval that contains a, $b \in \mathbb{R}$, it holds that*

$$\begin{aligned} f(b) &= f(a) + (b-a)f'(a) + \frac{(b-a)^2}{2!}f''(a) + \cdots + \frac{(b-a)^{n-1}}{(n-1)!}f^{(n-1)}(a) \\ &\quad + \int_a^b \frac{(b-s)^{n-1}}{(n-1)!}f^{(n)}(s)ds \\ &= \sum_{k=0}^{n-1} \frac{(b-a)^k}{k!}f^{(k)}(a) + \int_a^b \frac{(b-s)^{n-1}}{(n-1)!}f^{(n)}(s)ds. \end{aligned} \tag{3.30}$$

Here $f^{(k)}$ denotes the k-th derivative of f and $f^{(0)} := f$.

Proof. The basic relation of differentiation and integration[†6]

$$f(b) = f(a) + \int_a^b f'(s)ds$$

[6]At high school, this is the definition of definite integral, but at university, definite integral is defined in another way, and this formula is proved as 'the fundamental theorem of calculus'.

is Taylor's formula for $n = 1$. Applying integration by parts, we obtain

$$\begin{aligned} f(b) &= f(a) + \int_a^b (-(b-s)')f'(s)ds \\ &= f(a) + [-(b-s)f'(s)]_{s=a}^{s=b} - \int_a^b (-(b-s))\, f''(s)ds \\ &= f(a) + (b-a)f'(a) + \int_a^b (b-s)f''(s)ds. \end{aligned} \tag{3.31}$$

This is Taylor's formula for $n = 2$. Applying integration by parts again, we obtain

$$\begin{aligned} \int_a^b (b-s)f''(s)ds &= \int_a^b \left(-\frac{(b-s)^2}{2}\right)' f''(s)ds \\ &= \left[-\frac{(b-s)^2}{2}f''(s)\right]_{s=a}^{s=b} + \int_a^b \frac{(b-s)^2}{2}f'''(s)ds \\ &= \frac{(b-a)^2}{2}f''(a) + \int_a^b \frac{(b-s)^2}{2}f'''(s)ds. \end{aligned}$$

Substitute this for (3.31), and we obtain Taylor's formula for $n = 3$:

$$f(b) \ = \ f(a) + (b-a)f'(a) + \frac{(b-a)^2}{2}f''(a) + \int_a^b \frac{(b-s)^2}{2}f'''(s)ds. \tag{3.32}$$

Repeating this procedure will complete the proof. □

Example 3.7. (i) Applying Taylor's formula (3.32) to $f(x) = \log(1+x)$, and putting $a = 0$ and $b = x$, we obtain

$$\log(1+x) \ = \ x - \frac{1}{2}x^2 + \int_0^x \frac{(x-s)^2}{(1+s)^3}ds, \quad -1 < x. \tag{3.33}$$

Modifying this a little, for $a > 0$, we have,

$$\log(a+x) \ = \ \log a + \frac{x}{a} - \frac{1}{2}\left(\frac{x}{a}\right)^2 + \int_0^{\frac{x}{a}} \frac{\left(\frac{x}{a}-s\right)^2}{(1+s)^3}ds, \quad -a < x. \tag{3.34}$$

(ii) Applying Taylor's formula (3.32) to $f(x) = e^x$, and putting $a = 0$, $b = x$, we obtain

$$e^x \ = \ 1 + x + \frac{1}{2}x^2 + \int_0^x \frac{(x-s)^2}{2}e^s ds, \quad x \in \mathbb{R}. \tag{3.35}$$

In order to study a general function f, it is often very effective to approximate it by a quadratic function g, and study g in detail with the knowledge we learned at high school. Indeed, Taylor's formula (3.32) is useful for this purpose. For example, if $|x| \ll 1$, the integral terms in (3.33) and (3.35) are very small, so that the following approximation formulas hold.

$$\log(1+x) \approx x - \frac{1}{2}x^2,$$
$$e^x \approx 1 + x + \frac{1}{2}x^2.$$

In this sense, the integral term of Taylor's formula (3.30) is called the *remainder term.*†[7]

The remainder term often becomes small as $n \to \infty$. For example, in the formula

$$e^x = 1 + x + \frac{x^2}{2!} + \cdots + \frac{x^n}{n!} + \int_0^x \frac{(x-s)^n}{n!} e^s ds, \quad x \in \mathbb{R},$$

the remainder term converges to 0 for all $x \in \mathbb{R}$ as $n \to \infty$:

$$\begin{aligned} \left| \int_0^x \frac{(x-s)^n}{n!} e^s ds \right| &\leqq \left| \int_0^x \frac{|x-s|^n}{n!} \max\{e^x, e^0\} ds \right| \\ &\leqq \max\{e^x, 1\} \left| \int_0^x \frac{|x|^n}{n!} ds \right| \\ &= \max\{e^x, 1\} |x| \frac{|x|^n}{n!} \to 0, \quad n \to \infty. \end{aligned}$$

For the last convergence, see Proposition A.4. In other words, for all $x \in \mathbb{R}$, we have

$$1 + x + \frac{x^2}{2!} + \cdots + \frac{x^n}{n!} \to e^x, \quad n \to \infty.$$

Remark 3.6. In what follows, inequalities as we saw in the previous paragraph will appear so often. Basically, those are applications of the following inequality.

$$\left| \int_A^B f(t)dt \right| \leqq \left| \int_A^B |f(t)|\, dt \right| \leqq |A - B| \times \max_{\min\{A,B\} \leqq t \leqq \max\{A,B\}} |f(t)|.$$

Theorem 3.8. (Stirling's formula) *As* $n \to \infty$,

$$n! \sim \sqrt{2\pi}\, n^{n+\frac{1}{2}} e^{-n}. \tag{3.36}$$

Here '$\sim$' means that the ratio of the both sides converges to 1.

[7] The remainder term can also be described by high order derivatives, which is usually taught in first year at university.

Example 3.8. Applying Stirling's formula to the calculation of 10000!, we obtain an approximate value of it:

$$\sqrt{2\pi}\,10000^{10000+\frac{1}{2}}e^{-10000} = 2.84623596219 \times 10^{35659}.$$

The true value is $2.84625968091 \times 10^{35659}$, and hence the ratio (approximate value)/ (true value)= 0.999992.

Stirling's formula is necessary to estimate the binomial distribution precisely. For example, *Wallis' formula* follows immediately from it.

Corollary 3.1.

$$\lim_{n\to\infty}\binom{2n}{n}2^{-2n}\cdot\frac{1}{2}\sqrt{2n} = \frac{1}{\sqrt{2\pi}}.$$

Proof. By Stirling's formula, as $n\to\infty$, we have

$$\begin{aligned}\binom{2n}{n}2^{-2n}\cdot\frac{1}{2}\sqrt{2n} &= \frac{(2n)!}{(n!)^2}\cdot 2^{-2n}\cdot\frac{1}{2}\sqrt{2n}\\ &\sim \frac{\sqrt{2\pi}\,(2n)^{2n+\frac{1}{2}}e^{-2n}}{\left(\sqrt{2\pi}\,n^{n+\frac{1}{2}}e^{-n}\right)^2}\cdot 2^{-2n}\cdot\frac{1}{2}\sqrt{2n}\\ &= \frac{1}{\sqrt{2\pi}}.\end{aligned}$$

□

We prove Stirling's formula by Laplace's method ([Lapalce (1812)]).

Lemma 3.2. (Euler's integral (of the second kind))

$$n! = \int_0^\infty x^n e^{-x}\,dx,^{\dagger 8} \quad n\in\mathbb{N}_+. \tag{3.37}$$

Proof. We show (3.37) by mathematical induction. First, we show it for $n=1$. For $R>0$, applying integration by parts, we obtain

$$\begin{aligned}\int_0^R xe^{-x}dx &= \left[x(-e^{-x})\right]_0^R - \int_0^R (x)'(-e^{-x})dx\\ &= -Re^{-R} + \int_0^R e^{-x}dx\\ &= -Re^{-R} + \left[-e^{-x}\right]_0^R\\ &= -Re^{-R} - e^{-R} + 1.\end{aligned}$$

[8]Note that the improper integral in the right-hand side of (3.37) makes sense in case n is not an integer. $\Gamma(s+1) := \int_0^\infty x^s e^{-x}dx$ is called *Gamma function*, which defines 'factorial' for any real $s>0$.

By Proposition A.3 (i), the right-hand side of the last line converges to 1 as $R \to \infty$. Thus

$$\int_0^\infty xe^{-x}dx = 1,$$

which completes the proof of (3.37) for $n = 1$.

Next, assuming that (3.37) is valid for $n = k$, we show that it is also valid for $n = k + 1$. For $R > 0$, applying integration by parts, we obtain

$$\begin{aligned}\int_0^R x^{k+1}e^{-x}dx &= \left[x^{k+1}(-e^{-x})\right]_0^R - \int_0^R (k+1)x^k(-e^{-x})dx \\ &= -R^{k+1}e^{-R} + (k+1)\int_0^R x^k e^{-x}dx.\end{aligned}$$

Again, by Proposition A.3 (i), the first term of the last line converges to 0 as $R \to \infty$, while the second term converges to $(k+1)k! = (k+1)!$ by the assumption of the induction. Thus we have

$$\int_0^\infty x^{k+1}e^{-x}dx = (k+1)!\,,$$

which completes the proof. □

At first glance, $n!$ looks simpler than Euler's integral (3.37), but to study it in detail, the integral is much more useful.

Now, let us look closely at Euler's integral.

$$\int_0^\infty x^n e^{-x}dx = \int_0^\infty \exp\left(n\log x - x\right)dx = \int_0^\infty \exp\left(n\left(\log x - \frac{x}{n}\right)\right)dx.$$

The change of variables $t = x/n$ leads to

$$n! = n^{n+1}\int_0^\infty \exp\left(n\left(\log t - t\right)\right)dt. \tag{3.38}$$

Then, we know

$$f(t) := \log t - t, \quad t > 0,$$

is the key function.

We differentiate f to find an extremal value. The equation

$$f'(t) = \frac{1}{t} - 1 = 0$$

has a unique solution $t = 1$. Since $f''(t) = -1/t^2 < 0$, the extremal value $f(1) = -1$ is the maximum value of f (Fig. 3.6).

The following lemma is really amazing.

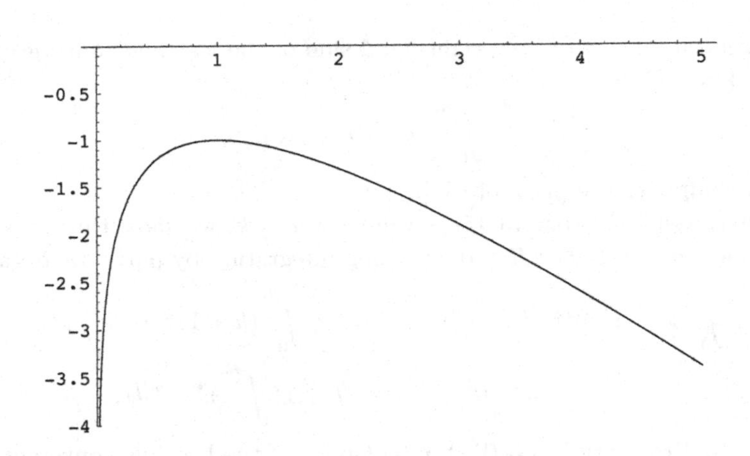

Fig. 3.6 The graph of $f(t)$

Lemma 3.3. *For any $0 < \delta < 1$ (Remark 3.1), it holds that*

$$\int_0^\infty \exp(nf(t))\,dt \sim \int_{1-\delta}^{1+\delta} \exp(nf(t))\,dt, \quad n \to \infty.$$

Namely, if $n \gg 1$, the integral of $\exp(nf(t))$ on $[0,\infty)$ is almost determined by the integral on arbitrarily small neighborhood of $t=1$, at which f takes the maximum value.

Proof. We define a function g by

$$g(t) := \begin{cases} \dfrac{-1-f(1-\delta)}{\delta}(t-1)-1 \ (0 < t \leqq 1), \\ \dfrac{f(1+\delta)+1}{\delta}(t-1)-1 \quad (1 < t). \end{cases}$$

The graph of g is a polyline that passes the following three points (Fig. 3.7): $(1-\delta, f(1-\delta))$, $(1,-1)$ and $(1+\delta, f(1+\delta))$.

Since f is a concave function, we have

$$g(t) \begin{cases} > f(t) \ (0 < t < 1-\delta), \\ \leqq f(t) \ (1-\delta \leqq t \leqq 1+\delta), \\ > f(t) \ (1+\delta < t). \end{cases}$$

Therefore

$$\frac{\displaystyle\int_{t>0\,;\,|t-1|>\delta} \exp(nf(t))\,dt}{\displaystyle\int_{1-\delta}^{1+\delta} \exp(nf(t))\,dt} \leqq \frac{\displaystyle\int_{t>0\,;\,|t-1|>\delta} \exp(ng(t))\,dt}{\displaystyle\int_{1-\delta}^{1+\delta} \exp(ng(t))\,dt}, \tag{3.39}$$

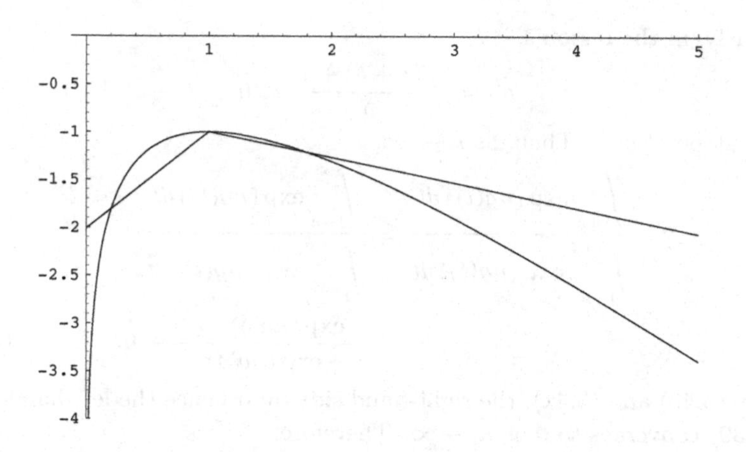

Fig. 3.7 The graph of $g(t)$ ($\delta = 4/5$, polyline)

where $\displaystyle\int_{t>0\,;\,|t-1|>\delta}$ denotes the integral $\displaystyle\int_0^{1-\delta} + \int_{1+\delta}^{\infty}$ on the set $\{t > 0\} \cap \{|t-1| > \delta\}$.

Since g is a polyline, we can explicitly calculate the right-hand side of (3.39). First, in the region $0 < t \leqq 1$,

$$a := \frac{-1 - f(1-\delta)}{\delta} > 0$$

is the slope of $g(t)$. Then, as $n \to \infty$,

$$\begin{aligned}
\frac{\displaystyle\int_0^{1-\delta} \exp(ng(t))\,dt}{\displaystyle\int_{1-\delta}^{1+\delta} \exp(ng(t))\,dt} &\leqq \frac{\displaystyle\int_{-\infty}^{1-\delta} \exp(ng(t))\,dt}{\displaystyle\int_{1-\delta}^{1} \exp(ng(t))\,dt} \\
&= \frac{\displaystyle\int_{-\infty}^{1-\delta} \exp(n(a(t-1)-1))\,dt}{\displaystyle\int_{1-\delta}^{1} \exp(n(a(t-1)-1))\,dt} \\
&= \frac{\dfrac{1}{na} e^{-n} \exp(-na\delta)}{\dfrac{1}{na} e^{-n} \left(1 - \exp(-na\delta)\right)} \\
&= \frac{\exp(-na\delta)}{1 - \exp(-na\delta)} \longrightarrow 0. \qquad (3.40)
\end{aligned}$$

Similarly, in the region $1 < t$,

$$a' := \frac{f(1+\delta)+1}{\delta} < 0$$

is the slope of $g(t)$. Then, as $n \to \infty$,

$$\frac{\displaystyle\int_{1+\delta}^{\infty} \exp(ng(t))\,dt}{\displaystyle\int_{1-\delta}^{1+\delta} \exp(ng(t))\,dt} \le \frac{\displaystyle\int_{1+\delta}^{\infty} \exp(ng(t))\,dt}{\displaystyle\int_{1}^{1+\delta} \exp(ng(t))\,dt} = \frac{\exp(na'\delta)}{1-\exp(na'\delta)} \longrightarrow 0. \tag{3.41}$$

By (3.40) and (3.41), the right-hand side (and hence the left-hand side) of (3.39) converges to 0 as $n \to \infty$. Therefore

$$\frac{\displaystyle\int_0^{\infty} \exp(nf(t))\,dt}{\displaystyle\int_{1-\delta}^{1+\delta} \exp(nf(t))\,dt} = \frac{\displaystyle\int_{1-\delta}^{1+\delta} \exp(nf(t))\,dt + \int_{t>0\,;\,|t-1|>\delta} \exp(nf(t))\,dt}{\displaystyle\int_{1-\delta}^{1+\delta} \exp(nf(t))\,dt}$$

$$= 1 + \frac{\displaystyle\int_{t>0\,;\,|t-1|>\delta} \exp(nf(t))\,dt}{\displaystyle\int_{1-\delta}^{1+\delta} \exp(nf(t))\,dt} \longrightarrow 1, \quad n \to \infty.$$

Thus the proof of Lemma 3.3 is complete. □

From (3.38) and Lemma 3.3, it follows that for any $0 < \delta < 1$,

$$n! \sim n^{n+1} \int_{1-\delta}^{1+\delta} \exp(nf(t))\,dt, \quad n \to \infty. \tag{3.42}$$

Furthermore, looking carefully at the above proof, if we replace δ by a positive decreasing sequence $\{\delta(n)\}_{n=1}^{\infty}$ converging to 0, there is still a possibility for (3.42) to hold. Indeed, for example,

$$\delta(n) := n^{-1/4}, \quad n = 2, 3, \ldots$$

satisfies the following.

$$n! \sim n^{n+1} \int_{1-\delta(n)}^{1+\delta(n)} \exp(nf(t))\,dt, \quad n \to \infty. \tag{3.43}$$

Let us prove (3.43). To this end, considering (3.40) and (3.41), it is enough to show

$$\lim_{n\to\infty} \exp(-na\delta(n)) = 0, \quad \lim_{n\to\infty} \exp(na'\delta(n)) = 0.$$

To show these two, considering the definitions of a and a', it is enough to show

$$\lim_{n\to\infty} n\,(1+f(1-\delta(n))) \;=\; -\infty, \qquad \lim_{n\to\infty} n\,(1+f(1+\delta(n))) \;=\; -\infty.$$

Since $f(t) = \log t - t$ and $\delta(n) = n^{-1/4}$, what we have to show are

$$\lim_{n\to\infty} n\left(\log(1-n^{-1/4}) + n^{-1/4}\right) = -\infty, \tag{3.44}$$

$$\lim_{n\to\infty} n\left(\log(1+n^{-1/4}) - n^{-1/4}\right) = -\infty. \tag{3.45}$$

First, as for (3.44), recalling (3.33), we see

$$\begin{aligned} n & \left(\log(1-n^{-1/4}) + n^{-1/4}\right) \\ &= n\left(-\frac{1}{2}\left(-n^{-1/4}\right)^2 - \int_{-n^{-1/4}}^{0} \frac{\left(-n^{-1/4}-s\right)^2}{(1+s)^3}ds\right) \\ &< -\frac{1}{2}n^{1/2} \;\to -\infty, \quad n\to\infty. \end{aligned}$$

Next, as for (3.45), we see

$$\begin{aligned} n & \left(\log(1+n^{-1/4}) - n^{-1/4}\right) \\ &= n\left(-\frac{1}{2}\left(n^{-1/4}\right)^2 + \int_{0}^{n^{-1/4}} \frac{\left(n^{-1/4}-s\right)^2}{(1+s)^3}ds\right) \\ &\leqq n\left(-\frac{1}{2}\left(n^{-1/4}\right)^2 + \int_{0}^{n^{-1/4}} \frac{\left(n^{-1/4}\right)^2}{1}ds\right) \\ &= -\frac{1}{2}n^{1/2} + n^{1/4} \;\to -\infty, \quad n\to\infty. \end{aligned}$$

From these (3.43) follows.

Applying Taylor's formula to the function $f(t)$, since $f(1) = -1$, $f'(1) = 0$, and $f''(s) = -1/s^2$, we have

$$\begin{aligned} f(t) &= f(1) + (t-1)f'(1) + \int_1^t (t-s)f''(s)ds \\ &= -1 - \int_1^t \frac{t-s}{s^2}ds. \end{aligned}$$

By (3.43), to prove Stirling's formula, it is enough to look at the behavior of $f(t)$ in the region $1-\delta(n) \leqq t \leqq 1+\delta(n)$. In this region, it holds that

$$-1 - \int_1^t \frac{t-s}{(1-\delta(n))^2}ds \;\leqq\; f(t) \;\leqq\; -1 - \int_1^t \frac{t-s}{(1+\delta(n))^2}ds,$$

namely,

$$-1-\frac{(t-1)^2}{2(1-\delta(n))^2} \leqq f(t) \leqq -1-\frac{(t-1)^2}{2(1+\delta(n))^2}.$$

If we put $b_\pm(n) := \sqrt{2}(1 \pm \delta(n))$, then $b_\pm(n) \to \sqrt{2}$ as $n \to \infty$, and

$$e^{-n}\int_{1-\delta(n)}^{1+\delta(n)} \exp\left(-\frac{n(t-1)^2}{b_-(n)^2}\right) dt$$
$$\leqq \int_{1-\delta(n)}^{1+\delta(n)} e^{nf(t)}dt \leqq e^{-n}\int_{1-\delta(n)}^{1+\delta(n)} \exp\left(-\frac{n(t-1)^2}{b_+(n)^2}\right) dt.$$

Multiplying all by e^n,

$$\int_{1-\delta(n)}^{1+\delta(n)} \exp\left(-\frac{n(t-1)^2}{b_-(n)^2}\right) dt$$
$$\leqq e^n\int_{1-\delta(n)}^{1+\delta(n)} e^{nf(t)}dt \leqq \int_{1-\delta(n)}^{1+\delta(n)} \exp\left(-\frac{n(t-1)^2}{b_+(n)^2}\right) dt. \tag{3.46}$$

Changing variables $u = \sqrt{n}(t-1)/b_\pm(n)$ leads to

$$\sqrt{n}\int_{1-\delta(n)}^{1+\delta(n)} \exp\left(-n\left(\frac{t-1}{b_\pm(n)}\right)^2\right) dt = b_\pm(n)\int_{-\sqrt{n}\delta(n)/b_\pm(n)}^{\sqrt{n}\delta(n)/b_\pm(n)} e^{-u^2}du.$$

Here, as $n \to \infty$, $\sqrt{n}\delta(n)/b_\pm(n) = n^{1/4}/b_\pm(n) \to \infty$, and hence the right-hand side of the above is convergent to (Remark 3.7)

$$\longrightarrow \sqrt{2}\int_{-\infty}^{\infty} e^{-u^2}du = \sqrt{2\pi}, \quad n \to \infty.$$

From this and (3.46), by the squeeze theorem, it follows that

$$\sqrt{n}e^n\int_{1-\delta(n)}^{1+\delta(n)} e^{nf(t)}dt \to \sqrt{2\pi}, \quad n \to \infty.$$

Therefore (3.43) implies

$$\sqrt{n}e^n\frac{n!}{n^{n+1}} \to \sqrt{2\pi}, \quad n \to \infty,$$

from which Stirling's formula (3.36) immediately follows. □

Remark 3.7. An improper integral

$$\int_{-\infty}^{\infty} \exp\left(-u^2\right) du = \sqrt{\pi}$$

or its equivalent

$$\int_{-\infty}^{\infty} \frac{1}{\sqrt{2\pi}} \exp\left(-\frac{u^2}{2}\right) du = 1$$

is called *Gaussian integral*, which appears very often in mathematics, physics, etc.

3.3.4 *Proof of de Moivre–Laplace's theorem*

The following is the key lemma to prove de Moivre–Laplace's theorem.

Lemma 3.4. *Let A, B $(A < B)$ be any two real numbers. Suppose that $\mathbb{N} \ni n, k \to \infty$ under the condition*

$$\frac{1}{2}n + \frac{A}{2}\sqrt{n} \leqq k \leqq \frac{1}{2}n + \frac{B}{2}\sqrt{n}. \tag{3.47}$$

Then, we have

$$b(k;n) := \binom{n}{k} 2^{-n} \sim \frac{1}{\sqrt{\frac{1}{2}\pi n}} \exp\left(-\frac{(k - \frac{1}{2}n)^2}{\frac{1}{2}n}\right).$$

More precisely, if we put

$$b(k;n) = \frac{1}{\sqrt{\frac{1}{2}\pi n}} \exp\left(-\frac{(k - \frac{1}{2}n)^2}{\frac{1}{2}n}\right)(1 + r_n(k)), \tag{3.48}$$

then $r_n(k)$ satisfies the following.

$$\max_{k\,;\,(3.47)} |r_n(k)| \to 0, \quad n \to \infty, \tag{3.49}$$

where $\max\limits_{k\,;\,(3.47)}$ denotes the maximum value over k satisfying (3.47) with n fixed.

Proof. [†9] If we put

$$n! = \sqrt{2\pi} n^{n+\frac{1}{2}} e^{-n}(1 + \eta_n), \tag{3.50}$$

we have $\eta_n \to 0$, as $n \to \infty$, by Stirling's formula (3.36). The following holds (Proposition A.2).

$$\max_{k\,;\,(3.47)} |\eta_k| \to 0, \quad n \to \infty. \tag{3.51}$$

Substituting (3.50) for $b(k;n) = \binom{n}{k} 2^{-n} = n!/((n-k)!k!) \cdot 2^{-n}$, we have

$$b(k;n) = \frac{1}{\sqrt{2\pi n \cdot \frac{k}{n}\left(\frac{n-k}{n}\right)}} \left(\frac{k}{n}\right)^{-k} \left(\frac{n-k}{n}\right)^{-n+k} 2^{-n} \frac{1 + \eta_n}{(1 + \eta_k)(1 + \eta_{n-k})}.$$

As $n \to \infty$,

$$\max_{k\,;\,(3.47)} \left|\frac{k}{n} - \frac{1}{2}\right| = \max_{k\,;\,(3.47)} \left|\frac{n-k}{n} - \frac{1}{2}\right| = \frac{\max\{|A|, |B|\}}{2\sqrt{n}} \to 0. \tag{3.52}$$

---

[9] The proof of this lemma (cf. [Sinai (1992)] Theorem 3.1) is difficult. Readers should challenge it after reading Sec. A.3.

Then, putting

$$b(k;n) = \frac{1}{\sqrt{\frac{1}{2}\pi n}} \left(\frac{k}{n}\right)^{-k} \left(\frac{n-k}{n}\right)^{-n+k} 2^{-n}(1+r_{n,k}),$$

by (3.51) and (3.52), we have† [10]

$$\max_{k\,;\,(3.47)} |r_{n,k}| \to 0, \quad n \to \infty. \tag{3.53}$$

Using the exponential function, we have

$$b(k;n) = \frac{1}{\sqrt{\frac{1}{2}\pi n}} \exp\left(T_{n,k}\right)\left(1+r_{n,k}\right),$$

where

$$T_{n,k} := -k \log \frac{k}{n} - (n-k)\log\frac{n-k}{n} + n \log \frac{1}{2}.$$

To show the lemma, putting

$$z := \frac{k - \frac{1}{2}n}{\frac{1}{2}\sqrt{n}},$$

we estimate the difference between $T_{n,k}$ and $-z^2/2$. First we write $T_{n,k}$ in terms of z,

$$T_{n,k} = -T^{(1)}_{n,k} - T^{(2)}_{n,k} + n \log \frac{1}{2},$$

where

$$T^{(1)}_{n,k} := k \log\left(\frac{1}{2} + \frac{1}{2}z\sqrt{\frac{1}{n}}\right), \quad T^{(2)}_{n,k} := (n-k)\log\left(\frac{1}{2} - \frac{1}{2}z\sqrt{\frac{1}{n}}\right).$$

Now, applying (3.34) with $a = \frac{1}{2}$ and $x = \frac{1}{2}z\sqrt{\frac{1}{n}}$, we see

$$T^{(1)}_{n,k} = k \log \frac{1}{2} + kz\sqrt{\frac{1}{n}} - \frac{k}{2}\cdot\frac{z^2}{n} + k\int_0^{z\sqrt{\frac{1}{n}}} \frac{\left(z\sqrt{\frac{1}{n}} - s\right)^2}{(1+s)^3} ds.$$

Let $\delta_{n,k}$ denote the last integral term (Remark 3.6). Then,

$$\begin{aligned} |\delta_{n,k}| &\leqq k \left| \int_0^{z\sqrt{\frac{1}{n}}} \frac{\left(z\sqrt{\frac{1}{n}}\right)^2}{\left(1 - \left|z\sqrt{\frac{1}{n}}\right|\right)^3} ds \right| = k\left|z\sqrt{\frac{1}{n}}\right|^3 \left(1 - \left|z\sqrt{\frac{1}{n}}\right|\right)^{-3} \\ &= \frac{k}{n}|z|^3\sqrt{\frac{1}{n}}\left(1 - \left|z\sqrt{\frac{1}{n}}\right|\right)^{-3}. \end{aligned} \tag{3.54}$$

[10] The proof of (3.53) needs the knowledge of continuity of functions of several variables. See Sec. A.3.3.

Under the condition (3.47), we have $|z| \leqq C := \max\{|A|, |B|\}$ so that

$$\max_{k\,;\,(3.47)} |\delta_{n,k}| \leqq \frac{\left(\frac{1}{2}n + \frac{1}{2}B\sqrt{n}\right)}{n} C^3 \sqrt{\frac{1}{n}} \left(1 - C\sqrt{\frac{1}{n}}\right)^{-3} \to 0, \quad n \to \infty.$$

Similarly, applying (3.34) with $a = \frac{1}{2}$ and $x = -\frac{1}{2}z\sqrt{\frac{1}{n}}$, we see

$$T_{n,k}^{(2)} = (n-k)\log\frac{1}{2} - (n-k)z\sqrt{\frac{1}{n}} - \frac{n-k}{2}\cdot\frac{z^2}{n} + (n-k)\int_0^{-z\sqrt{\frac{1}{n}}} \frac{\left(-z\sqrt{\frac{1}{n}} - s\right)^2}{(1+s)^3} ds.$$

Let $\delta'_{n,k}$ denote the last integral term. Then, we also have

$$\max_{k\,;\,(3.47)} |\delta'_{n,k}| \to 0, \quad n \to \infty.$$

Putting $\delta''_{n,k} := -\delta_{n,k} - \delta'_{n,k}$, it holds that

$$\begin{aligned} T_{n,k} &= -\left(k\log\frac{1}{2} + kz\sqrt{\frac{1}{n}} - \frac{k}{2}\cdot\frac{z^2}{n} + \delta_{n,k}\right) \\ &\quad - \left((n-k)\log\frac{1}{2} - (n-k)z\sqrt{\frac{1}{n}} - \frac{n-k}{2}\cdot\frac{z^2}{n} + \delta'_{n,k}\right) + n\log\frac{1}{2} \\ &= (n-2k)z\sqrt{\frac{1}{n}} + \frac{z^2}{2} + \delta''_{n,k}. \end{aligned}$$

Recall that $k = \frac{1}{2}z\sqrt{n} + \frac{1}{2}n$, and we see

$$T_{n,k} = -\frac{z^2}{2} + \delta''_{n,k}$$

with

$$\max_{k\,;\,(3.47)} |\delta''_{n,k}| \to 0, \quad n \to \infty. \tag{3.55}$$

From all the above,

$$\begin{aligned} b(k;n) &= \frac{1}{\sqrt{\frac{1}{2}\pi n}} \exp\left(T_{n,k}\right)(1 + r_{n,k}) \\ &= \frac{1}{\sqrt{\frac{1}{2}\pi n}} \exp\left(-\frac{z^2}{2} + \delta''_{n,k}\right)(1 + r_{n,k}) \\ &= \frac{1}{\sqrt{\frac{1}{2}\pi n}} \exp\left(-\frac{(k - \frac{1}{2}n)^2}{\frac{1}{2}n} + \delta''_{n,k}\right)(1 + r_{n,k}). \end{aligned}$$

On the other hand, by (3.48),

$$r_n(k) = \frac{b(k;n)}{\dfrac{1}{\sqrt{\frac{1}{2}\pi n}}\exp\left(-\dfrac{(k-\frac{1}{2}n)^2}{\frac{1}{2}n}\right)} - 1$$

and hence

$$r_n(k) = \exp\left(\delta''_{n,k}\right)(1+r_{n,k}) - 1.$$

From this, (3.53) and (3.55), the desired (3.49) follows.†[11] □

Proof of Theorem 3.6. By Theorem 3.5 and Lemma 3.4,

$$\begin{aligned} P_n(A \leqq Z_n \leqq B) &= P_n\left(\frac{n}{2} + \frac{A}{2}\sqrt{n} \leqq \xi_1 + \cdots + \xi_n \leqq \frac{n}{2} + \frac{B}{2}\sqrt{n}\right) \\ &= \sum_{k\,;\,(3.47)} b(k;n) \\ &= \sum_{k\,;\,(3.47)} \frac{1}{\sqrt{\frac{1}{2}\pi n}}\exp\left(-\frac{(k-\frac{1}{2}n)^2}{\frac{1}{2}n}\right)(1+r_n(k)), \end{aligned}$$

where $\sum_{k\,;\,(3.47)}$ denotes the sum over k satisfying the condition (3.47) with n fixed. For each $k \in \mathbb{N}_+$, let

$$z_k := \frac{k-\frac{1}{2}n}{\frac{1}{2}\sqrt{n}}.$$

Noting that the length between two adjacent z_k and z_{k+1} is $1/\left(\frac{1}{2}\sqrt{n}\right)$, we rewrite $P_n(A \leqq Z_n \leqq B)$ as a Riemann sum and a remainder:

$$\begin{aligned} P_n(A \leqq Z_n \leqq B) &= \frac{1}{\frac{1}{2}\sqrt{n}} \sum_{A \leqq z_k \leqq B} \frac{1}{\sqrt{2\pi}}\exp\left(-\frac{z_k^2}{2}\right)(1+r_n(k)) \\ &= \frac{1}{\frac{1}{2}\sqrt{n}} \sum_{A \leqq z_k \leqq B} \frac{1}{\sqrt{2\pi}}\exp\left(-\frac{z_k^2}{2}\right) \\ &\quad + \frac{1}{\frac{1}{2}\sqrt{n}} \sum_{A \leqq z_k \leqq B} \frac{1}{\sqrt{2\pi}}\exp\left(-\frac{z_k^2}{2}\right) r_n(k). \end{aligned}$$

The limit of the first term, the Riemann sum, is

$$\lim_{n\to\infty} \frac{1}{\frac{1}{2}\sqrt{n}} \sum_{A \leqq z_k \leqq B} \frac{1}{\sqrt{2\pi}}\exp\left(-\frac{z_k^2}{2}\right) = \int_A^B \frac{1}{\sqrt{2\pi}}\exp\left(-\frac{z^2}{2}\right)dz.$$

[11] The proof of (3.49) needs the knowledge of continuous function of several variables. See Sec. A.3.3.

As for the second term, since the absolute value of the first term $\left|\frac{1}{\frac{1}{2}\sqrt{n}}\sum_{A \leqq z_k \leqq B}\cdots\right|$ is bounded by a constant $M > 0$ that is independent of n (Proposition A.1), we see

$$\left|\frac{1}{\frac{1}{2}\sqrt{n}} \sum_{A \leqq z_k \leqq B} \frac{1}{\sqrt{2\pi}} \exp\left(-\frac{z_k^2}{2}\right) r_n(k)\right| \leqq M \times \max_{k\,;\,(3.47)} |r_n(k)| \to 0, \quad n \to \infty.$$

This completes the proof of de Moivre–Laplace's theorem. □

Example 3.9. Applying de Moivre–Laplace's theorem, let us calculate an approximate value of the following probability.

$$P_{100}\left(\xi_1 + \cdots + \xi_{100} \geqq 55\right) = \sum_{k=55}^{100} \binom{100}{k} 2^{-100}. \tag{3.56}$$

This is the probability that the number of Heads is more than or equal to 55 among 100 coin tosses. Although the random variable $\xi_1 + \cdots + \xi_{100}$ takes only integers $0, \ldots, 100$, just as we considered the event $\{\xi_1 + \cdots + \xi_{100} = k\}$ as $\{k-(1/2) \leqq \xi_1 + \cdots + \xi_{100} < k+(1/2)\}$ in the histogram of Fig. 3.3(right), if we consider (3.56) as

$$P_{100}\left(\xi_1 + \cdots + \xi_{100} \geqq 54.5\right),$$

and apply de Moivre–Laplace's theorem, the accuracy of approximation will be improved. This is called the *continuity correction*. By this method, we obtain

$$\begin{aligned} P_{100}\left(\xi_1 + \cdots + \xi_{100} \geqq 54.5\right) &= P_{100}\left(\frac{\xi_1 + \cdots + \xi_{100} - 50}{\frac{1}{2}\sqrt{100}} \geqq \frac{54.5 - 50}{\frac{1}{2}\sqrt{100}}\right) \\ &= P_{100}\left(\frac{\xi_1 + \cdots + \xi_{100} - 50}{\frac{1}{2}\sqrt{100}} \geqq 0.9\right) \\ &\approx \int_{0.9}^{\infty} \frac{1}{\sqrt{2\pi}} \exp\left(-\frac{x^2}{2}\right) dx = 0.18406. \end{aligned}$$

Since the true value of (3.56) is

$$\begin{aligned} \sum_{k=55}^{100} \binom{100}{k} 2^{-100} &= 233375500604595657604761955760 \times 2^{-100} \\ &= 0.1841008087, \end{aligned}$$

the approximation error is only 0.00004.

Example 3.10. After making a continuity correction, we applied de Moivre–Laplace's theorem to the following case.

$$P_{10^8}\left(\left|\frac{\xi_1+\cdots+\xi_{10^8}}{10^8}-\frac{1}{2}\right| \geqq \frac{1}{2000}\right)$$
$$\approx 2\int_{9.9999}^{\infty}\frac{1}{\sqrt{2\pi}}\exp\left(-\frac{x^2}{2}\right)dx \;=\; 1.52551\times 10^{-23}.$$

This is even more precise than the inequality (3.22). Recalling that the Avogadro constant is 6.02×10^{23} helps us imagine how small this probability is. The same probability was estimated as "$\leqq 1/100$" by Chebyshev's inequality (3.15). This fact tells us Chebyshev's inequality gives very loose bounds. However, do not jump to the conclusion that it is therefore a bad inequality. Rather, we should admire Chebyshev's great insight that realized this loose inequality is enough to prove the law of large numbers.

Do readers remember that the inequality (3.2) used in the proof of Theorem 3.1 was also very loose?

3.4 Central limit theorem

The assertion of de Moivre–Laplace's theorem is valid not only for coin tosses but also for general sequences of i.i.d. random variables.

Suppose that for each $n\in\mathbb{N}_+$, a probability space $(\Omega_n,\mathfrak{P}(\Omega_n),\mu_n)$, and i.i.d. random variables $X_{n,1},X_{n,2},\ldots,X_{n,n}$ are given. Let their common mean be

$$\mathbf{E}[X_{n,k}] \;=\; m_n.$$

In addition, suppose that there exist constants $\sigma^2>0$ and $R>0$ independent of n, such that

$$\mathbf{V}[X_{n,k}] = \sigma_n^2 \geqq \sigma^2, \tag{3.57}$$
$$\max_{\omega\in\Omega_n}|X_{n,k}(\omega)-m_n| \leqq R. \tag{3.58}$$

Consider the standardization of $X_{n,1}+\cdots+X_{n,n}$:

$$Z_n := \frac{(X_{n,1}-m_n)+\cdots+(X_{n,n}-m_n)}{\sqrt{\sigma_n^2 n}}. \tag{3.59}$$

Theorem 3.9. *For any $A<B$, it holds that*

$$\lim_{n\to\infty}\mu_n\left(A\leqq Z_n\leqq B\right) = \int_A^B\frac{1}{\sqrt{2\pi}}\exp\left(-\frac{x^2}{2}\right)dx.$$

Example 3.11. Let $0 < p < 1$ and let $(\{0,1\}^n, \mathfrak{P}(\{0,1\}^n), P_n^{(p)})$ be the probability space of Example 3.2. Then, the coordinate functions $\{\xi_k\}_{k=1}^n$ is unfair coin tosses if $p \neq 1/2$. Since $\mathbf{E}[\xi_k] = p$ and $\mathbf{V}[\xi_k] = p(1-p)$, Theorem 3.9 implies

$$\lim_{n\to\infty} P_n^{(p)}\left(A \leqq \frac{(\xi_1 - p) + \cdots + (\xi_n - p)}{\sqrt{p(1-p)n}} \leqq B\right) = \int_A^B \frac{1}{\sqrt{2\pi}} e^{-x^2/2} dx.$$

In general, a theorem such as Theorem 3.9 that asserts "the distribution of the standardized sum of n random variables converges to the standard normal distribution as $n \to \infty$" is called the *central limit theorem.* Among numerous limit theorems, it is the most important and hence Polya named it 'central'. Proving Theorem 3.9 rigorously exceeds the level of this book, so that we here give a convincing explanation that it must be true.

As the proof of Theorem 3.5 somewhat implies, the moment generating function $M_X(t)$ of a random variable X determines the distribution of X.

Proposition 3.6. *Let X and Y be random variables. If $M_X(t) = M_Y(t)$, $t \in \mathbb{R}$, then X and Y are identically distributed.*

Proof. Let $a_1 < \cdots < a_s$ be all possible values that X takes. We have

$$\begin{aligned}
&\lim_{t\to\infty} \frac{1}{t} \log M_X(t) \\
&= \lim_{t\to\infty} \frac{1}{t} \log \sum_{i=1}^{s} \exp(ta_i) P(X = a_i) \\
&= \lim_{t\to\infty} \frac{1}{t} \log \left(\exp(ta_s) \sum_{i=1}^{s} \exp(t(a_i - a_s)) P(X = a_i) \right) \\
&= \lim_{t\to\infty} \left(\frac{1}{t} \log \exp(ta_s) + \frac{1}{t} \log \sum_{i=1}^{s} \exp(t(a_i - a_s)) P(X = a_i) \right) \\
&= a_s + \lim_{t\to\infty} \frac{1}{t} \log \left(P(X = a_s) + \sum_{i=1}^{s-1} \exp(t(a_i - a_s)) P(X = a_i) \right) \\
&= a_s.
\end{aligned}$$

After knowing a_s,

$$\begin{aligned}\lim_{t\to\infty} M_X(t)\exp(-ta_s) &= \lim_{t\to\infty}\sum_{i=1}^{s}\exp(t(a_i-a_s))P(X=a_i)\\ &= P(X=a_s)+\lim_{t\to\infty}\sum_{i=1}^{s-1}\exp(t(a_i-a_s))P(X=a_i)\\ &= P(X=a_s).\end{aligned}$$

Thus we can obtain a_s and $P(X = a_s)$ from $M_X(t)$. If we apply this procedure to

$$M_X(t)-\exp(ta_s)P(X=a_s),$$

then we obtain a_{s-1} and $P(X = a_{s-1})$. Repeating this procedure, we obtain the distribution of X from $M_X(t)$. In particular, $M_X(t) = M_Y(t)$, $t \in \mathbb{R}$, implies that X and Y are identically distributed. □

Now, let us calculate the moment generating function of the standardized sum (3.24) of n coin tosses $\{\xi_i\}_{i=1}^n$, and observe its asymptotic behavior.

$$\begin{aligned}&\mathbf{E}\left[\exp\left(t\cdot\frac{(\xi_1-\frac12)+\cdots+(\xi_n-\frac12)}{\frac12\sqrt{n}}\right)\right]\\ &= \mathbf{E}\left[\exp\left(t\cdot\frac{\xi_1-\frac12}{\frac12\sqrt{n}}\right)\times\cdots\times\exp\left(t\cdot\frac{\xi_n-\frac12}{\frac12\sqrt{n}}\right)\right]\\ &= \mathbf{E}\left[\exp\left(t\cdot\frac{\xi_1-\frac12}{\frac12\sqrt{n}}\right)\right]\times\cdots\times\mathbf{E}\left[\exp\left(t\cdot\frac{\xi_n-\frac12}{\frac12\sqrt{n}}\right)\right]\\ &= \mathbf{E}\left[\exp\left(t\cdot\frac{\xi_1-\frac12}{\frac12\sqrt{n}}\right)\right]^n\\ &= \left(\frac12\cdot\exp\left(-\frac{t}{\sqrt{n}}\right)+\frac12\cdot\exp\left(\frac{t}{\sqrt{n}}\right)\right)^n\\ &= \left(1+\frac12\left(\exp\left(\frac{t}{2\sqrt{n}}\right)-\exp\left(-\frac{t}{2\sqrt{n}}\right)\right)^2\right)^n\\ &= \left(1+\frac{c_n(t)}{n}\right)^n, \qquad (3.60)\end{aligned}$$

where

$$c_n(t) := \frac{n}{2}\left(\exp\left(\frac{t}{2\sqrt{n}}\right)-\exp\left(-\frac{t}{2\sqrt{n}}\right)\right)^2 \to \frac{t^2}{2}, \quad n\to\infty. \qquad (3.61)$$

The last convergence can be proved in various ways. Here we prove it by Taylor's formula (3.31):

$$e^x = 1 + x + \int_0^x (x-s)e^s ds, \quad x \in \mathbb{R}.$$

According to this formula, if $|x| \ll 1$, we have $e^x \approx 1 + x$, and hence

$$c_n(t) \approx \frac{n}{2}\left(\left(1+\frac{t}{2\sqrt{n}}\right)-\left(1-\frac{t}{2\sqrt{n}}\right)\right)^2 = \frac{n}{2}\left(\frac{t}{\sqrt{n}}\right)^2 = \frac{t^2}{2}.$$

To prove (3.61), we have to show that the remainder term converges to 0 as $n \to \infty$, which can be done in the same way as (3.54).

Finally, from (3.60) and (3.61), and Lemma 3.5 below, it follows that for each $t \in \mathbb{R}$, we have

$$\mathbf{E}\left[\exp\left(t \cdot \frac{(\xi_1 - \frac{1}{2}) + \cdots + (\xi_n - \frac{1}{2})}{\frac{1}{2}\sqrt{n}}\right)\right] \to \exp\left(\frac{t^2}{2}\right), \quad n \to \infty. \quad (3.62)$$

Lemma 3.5. *If a sequence of real numbers* $\{c_n\}_{n=1}^{\infty}$ *converges to* $c \in \mathbb{R}$, *then*

$$\lim_{n\to\infty}\left(1+\frac{c_n}{n}\right)^n = e^c.$$

Proof. Since $c_n/n \to 0$, $n \to \infty$, we may assume $|c_n/n| < 1$ for $n \gg 1$. By (3.33), using the estimation that led to (3.54), we get the following.

$$\begin{aligned}
\left|n\log\left(1+\frac{c_n}{n}\right)-c\right| &\leqq \left|n\log\left(1+\frac{c_n}{n}\right)-c_n\right| + |c_n - c| \\
&= n\left|-\frac{1}{2}\left(\frac{c_n}{n}\right)^2 + \int_0^{c_n/n}\frac{\left(\frac{c_n}{n}-t\right)^2}{(1+t)^3}dt\right| + |c_n - c| \\
&\leqq \frac{c_n^2}{2n} + n\left|\int_0^{c_n/n}\frac{\left|\frac{c_n}{n}\right|^2}{\left(1-\left|\frac{c_n}{n}\right|\right)^3}dt\right| + |c_n - c| \\
&= \frac{c_n^2}{2n} + n\cdot\left|\frac{c_n}{n}\right|\frac{\left|\frac{c_n}{n}\right|^2}{\left(1-\left|\frac{c_n}{n}\right|\right)^3} + |c_n - c| \\
&= \frac{c_n^2}{2n} + |c_n|\frac{\left|\frac{c_n}{n}\right|^2}{\left(1-\left|\frac{c_n}{n}\right|\right)^3} + |c_n - c| \to 0, \quad n \to 0.
\end{aligned}$$

Therefore

$$\lim_{n\to\infty} n\log\left(1+\frac{c_n}{n}\right) = c.$$

By the continuity of the exponential function,

$$\begin{aligned}\lim_{n\to\infty}\left(1+\frac{c_n}{n}\right)^n &= \lim_{n\to\infty}\exp\left(n\log\left(1+\frac{c_n}{n}\right)\right)\\ &= \exp\left(\lim_{n\to\infty}n\log\left(1+\frac{c_n}{n}\right)\right) = e^c.\end{aligned}$$

□

The function $\exp(t^2/2)$, which is the limit in (3.62), has a deep relation with the density function (3.29) of the standard normal distribution.

Lemma 3.6. *For each* $t \in \mathbb{R}$,

$$\int_{-\infty}^{\infty} e^{tx}\cdot\frac{1}{\sqrt{2\pi}}\exp\left(-\frac{x^2}{2}\right)dx = \exp\left(\frac{t^2}{2}\right).$$

Proof. We complete the square for the quadratic function of x in the exponent:

$$\begin{aligned}&\int_{-\infty}^{\infty} e^{tx}\cdot\frac{1}{\sqrt{2\pi}}\exp\left(-\frac{x^2}{2}\right)dx\\ &= \int_{-\infty}^{\infty}\frac{1}{\sqrt{2\pi}}\exp\left(tx-\frac{1}{2}x^2\right)dx\\ &= \int_{-\infty}^{\infty}\frac{1}{\sqrt{2\pi}}\exp\left(-\frac{1}{2}(x-t)^2+\frac{1}{2}t^2\right)dx\\ &= \int_{-\infty}^{\infty}\frac{1}{\sqrt{2\pi}}\exp\left(-\frac{(x-t)^2}{2}\right)dx\cdot\exp\left(\frac{t^2}{2}\right)\\ &= \int_{-\infty}^{\infty}\frac{1}{\sqrt{2\pi}}\exp\left(-\frac{x^2}{2}\right)dx\cdot\exp\left(\frac{t^2}{2}\right)\\ &= \exp\left(\frac{t^2}{2}\right).\end{aligned}$$

The last '=' is due to Remark 3.7. □

Considering Proposition 3.6, the convergence (3.62), and Lemma 3.6, let us show, as a substitute of the proof of Theorem 3.9, that the moment generating function of Z_n defined by (3.59) satisfies

$$\lim_{n\to\infty} M_{Z_n}(t) = \exp\left(\frac{t^2}{2}\right), \quad t\in\mathbb{R}.$$

We calculate $M_{Z_n}(t)$ explicitly:

$$M_{Z_n}(t) = \mathbf{E}\left[\exp\left(t \cdot \frac{(X_{n,1} - m_n) + \cdots + (X_{n,n} - m_n)}{\sqrt{\sigma_n^2 n}}\right)\right]$$

$$= \mathbf{E}\left[\exp\left(t \cdot \frac{X_{n,1} - m_n}{\sqrt{\sigma_n^2 n}}\right)\right]^n.$$

By Taylor's formula (3.35), we see

$$\mathbf{E}\left[\exp\left(t \cdot \frac{X_{n,1} - m_n}{\sqrt{\sigma_n^2 n}}\right)\right]$$

$$= \mathbf{E}\left[1 + t \cdot \frac{X_{n,1} - m_n}{\sqrt{\sigma_n^2 n}} + \frac{t^2}{2} \cdot \frac{(X_{n,1} - m_n)^2}{\sigma_n^2 n} + \frac{r(n,t)}{n}\right]$$

$$= 1 + \frac{t^2}{2}\mathbf{E}\left[\frac{(X_{n,1} - m_n)^2}{\sigma_n^2 n}\right] + \mathbf{E}\left[\frac{r(n,t)}{n}\right]$$

$$= 1 + \frac{t^2}{2n} + \frac{\mathbf{E}\left[r(n,t)\right]}{n},$$

where the remainder term $r(n,t)$ is given by

$$r(n,t) = n \int_0^{t \cdot \frac{X_{n,1} - m_n}{\sqrt{\sigma_n^2 n}}} \frac{1}{2}\left(t \cdot \frac{X_{n,1} - m_n}{\sqrt{\sigma_n^2 n}} - s\right)^2 e^s ds,$$

whose mean is then estimated by (3.57) and (3.58) as

$$|\mathbf{E}\left[r(n,t)\right]|$$

$$\leqq \mathbf{E}\left[n \cdot \left|t \cdot \frac{X_{n,1} - m_n}{\sqrt{\sigma_n^2 n}}\right| \cdot \frac{1}{2} \cdot \left|t \cdot \frac{X_{n,1} - m_n}{\sqrt{\sigma_n^2 n}}\right|^2 \cdot \exp\left(\left|t \cdot \frac{X_{n,1} - m_n}{\sqrt{\sigma_n^2 n}}\right|\right)\right]$$

$$\leqq \mathbf{E}\left[n \cdot \frac{|t|^3}{2}\left|\frac{R}{\sqrt{\sigma^2 n}}\right|^3 \exp\left(\left|t \cdot \frac{R}{\sqrt{\sigma^2 n}}\right|\right)\right]$$

$$\leqq \frac{1}{2\sqrt{n}}|t|^3\left(\frac{R}{\sqrt{\sigma^2}}\right)^3 \exp\left(\left|t \cdot \frac{R}{\sqrt{\sigma^2 n}}\right|\right) \to 0, \quad n \to \infty.$$

Therefore

$$\mathbf{E}\left[\exp\left(t \cdot \frac{X_{n,1} - m_n}{\sqrt{\sigma_n^2 n}}\right)\right] = 1 + \frac{c_n(t)}{n},$$

$$c_n(t) = \frac{t^2}{2} + \mathbf{E}\left[r(n,t)\right] \to \frac{t^2}{2}, \quad n \to \infty.$$

Hence by Lemma 3.5, as $n \to \infty$, we see

$$M_{Z_n}(t) = \mathbf{E}\left[\exp\left(t \cdot \frac{X_{n,1} - m_n}{\sqrt{\sigma_n^2 n}}\right)\right]^n = \left(1 + \frac{c_n(t)}{n}\right)^n \to \exp\left(\frac{t^2}{2}\right).$$

□

3.5 Mathematical statistics

3.5.1 *Inference*

Let us look at applications of limit theorems in mathematical statistics. We begin with the *statistical inference*. It provides guidelines to construct stochastic models—probability spaces, random variables, etc., from given statistical data. Consider the following exercise.†[12]

> **Exercise II** 1,000 thumbtacks were thrown on a flat floor, and 400 of them landed point up. From this, estimate the probability that the thumbtack lands point up.

We consider Exercise II using the mathematical model of unfair coin tosses, i.e., the probability space $(\{0,1\}^n, \mathfrak{P}(\{0,1\}^n), P_n^{(p)})$ introduced in Example 3.2. The situation of Exercise II is interpreted that $n = 1000$, $0 < p < 1$ is the probability that we want to know, and that the coordinate function $\xi_i : \{0,1\}^{1000} \to \{0,1\}$ is the state (point up($=1$) or point down($=0$)) of the i-th thumbtack for each i. Then, a random variable

$$S := \xi_1 + \cdots + \xi_{1000}$$

is the number of thumbtacks that land point up. The experiment is interpreted as choice of an $\hat{\omega}$ from $\{0,1\}^{1000}$, and its outcome is interpreted as

$$S(\hat{\omega}) = \xi_1(\hat{\omega}) + \cdots + \xi_{1000}(\hat{\omega}) = 400.$$

Since $\mathbf{E}[S] = 1000p$ and $\mathbf{V}[S] = 1000p(1-p)$, Chebyshev's inequality implies

$$\begin{aligned} P_{1000}^{(p)}\left(\left| \frac{S}{1000} - p \right| \geqq \varepsilon \right) &= P_{1000}^{(p)}\left(|S - 1000p| \geqq 1000\varepsilon \right) \\ &\leqq \frac{\mathbf{V}[S]}{(1000\varepsilon)^2} = \frac{p(1-p)}{1000\varepsilon^2} \leqq \frac{1}{4000\varepsilon^2}. \end{aligned}$$

Here putting $\varepsilon := \sqrt{10}/20 = 0.158$, we obtain

$$P_{1000}^{(p)}\left(\left| \frac{S}{1000} - p \right| \geqq 0.158 \right) \leqq \frac{1}{100}.$$

Therefore

$$P_{1000}^{(p)}\left(\left| \frac{S}{1000} - p \right| < 0.158 \right) \geqq \frac{99}{100}.$$

[12] Here we deal with 'interval estimation of population proportion'.

Solving in p in $P_{1000}^{(p)}(\quad)$ of the left-hand side, we see the probability that

$$\frac{S}{1000} - 0.158 < p < \frac{S}{1000} + 0.158$$

is not less than 0.99. We now consider that $S(\hat{\omega}) = 400$ is an outcome of the occurrence of the above event. Then, we obtain

$$0.242 < p < 0.558.$$

This estimate is not so good because Chebyshev's inequality is loose.

The central limit theorem (Example 3.11) gives much more precise estimate:

$$P_{1000}^{(p)}\left(\left|\frac{S-1000p}{\sqrt{1000p(1-p)}}\right| \geqq z\right) \approx 2\int_z^{\infty} \frac{1}{\sqrt{2\pi}} \exp\left(-\frac{x^2}{2}\right) dx, \quad z > 0.$$

Now, for $0 < \alpha < 1/2$, let $z(\alpha)$ denote a positive real number z such that

$$2\int_z^{\infty} \frac{1}{\sqrt{2\pi}} \exp\left(-\frac{x^2}{2}\right) dx = \alpha.$$

$z(\alpha)$ is called the $100 \times \alpha\%$ point of the standard normal distribution (Fig. 3.8).

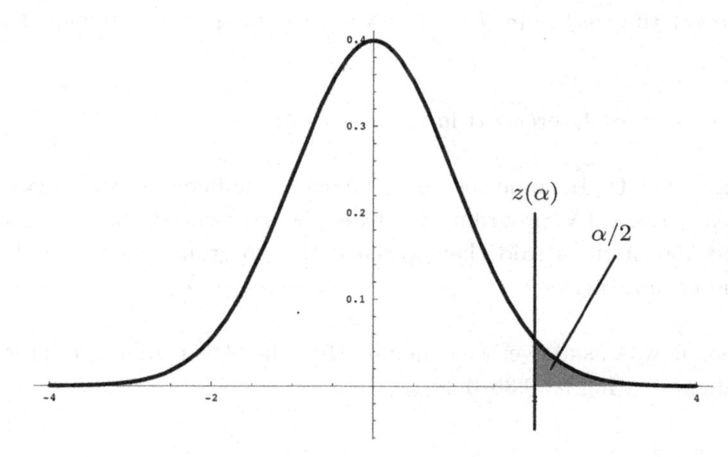

Fig. 3.8 The $100 \times \alpha\%$ point $z(\alpha)$ of the standard normal distribution

Now, we have

$$P_{1000}^{(p)}\left(\left|\frac{S-1000p}{\sqrt{1000p(1-p)}}\right| < z(\alpha)\right) \approx 1 - \alpha,$$

i.e., the probability that

$$\frac{S}{1000} - z(\alpha)\sqrt{\frac{p(1-p)}{1000}} < p < \frac{S}{1000} + z(\alpha)\sqrt{\frac{p(1-p)}{1000}}$$

is approximately $1-\alpha$. Let $0 < \alpha \ll 1$. As before, we consider that $S(\hat{\omega}) = 400$ is an outcome of the occurrence of the above event. Namely, we judge that

$$0.4 - z(\alpha)\sqrt{\frac{p(1-p)}{1000}} < p < 0.4 + z(\alpha)\sqrt{\frac{p(1-p)}{1000}}.$$

We have to solve the above inequality in p, but here, we approximate it by assuming $p \approx 0.4$ due to the law of large numbers:

$$0.4 - z(\alpha)\sqrt{\frac{0.4 \times (1-0.4)}{1000}} < p < 0.4 + z(\alpha)\sqrt{\frac{0.4 \times (1-0.4)}{1000}}.$$

For example, if $\alpha = 0.01$, then $z(0.01) = 2.58$, and hence the above inequality is now

$$0.36 < p < 0.44.$$

We express this result as "The 99% *confidence interval* of p is $[0.36, 0.44]$". The greater the *confidence level* $1-\alpha$ is, the wider the confidence interval becomes.

Let us present Exercise II in another form.

Exercise II′ In some region, to make an audience rating survey of a certain TV program, 1000 people were chosen at random, and 400 of them said they watched the program. Estimate the audience rating.

In the same way as above, we conclude that the 99% confidence interval of the audience rating is $[0.36, 0.44]$.

3.5.2 *Test*

The *statistical test* provides guidelines to judge whether or not given stochastic models—probability spaces, random variables, etc., do not contradict with observed statistical data. Consider the following exercise.†[13]

[13] Here we deal with 'test of population proportion'.

Exercise III A coin was tossed 200 times, and it came up Heads 115 times. Is it a fair coin?

First, we state a *hypothesis* H.

$$\mathrm{H}: \text{The coin is fair.}$$

Under the hypothesis H, we consider the probability space

$$(\{0,1\}^{200}, \mathfrak{P}(\{0,1\}^{200}), P_{200})$$

and the coordinate functions $\{\xi_i\}_{i=1}^{200}$. The number of Heads in 200 coin tosses $\omega \in \{0,1\}^{200}$ is $S(\omega) := \xi_1(\omega) + \cdots + \xi_{200}(\omega)$. Let us then calculate the probability

$$P_{200}\left(\,|S-100| \geqq 15\,\right).$$

After making a continuity correction, we apply de Moivre–Laplace's theorem:

$$\begin{aligned} P_{200}\left(\,|S-100| \geqq 15\,\right) &= P_{200}\left(\left|\frac{S-100}{\frac{1}{2}\sqrt{200}}\right| \geqq \frac{14.5}{\frac{1}{2}\sqrt{200}}\right) \\ &\approx 2\int_{14.5/(\frac{1}{2}\sqrt{200})}^{\infty} \frac{1}{\sqrt{2\pi}} \exp\left(-\frac{x^2}{2}\right) dx \\ &= 2\int_{2.05061}^{\infty} \frac{1}{\sqrt{2\pi}} \exp\left(-\frac{x^2}{2}\right) dx \\ &= 0.040305. \end{aligned}$$

Therefore, under the hypothesis H, the event of Exercise III is of probability 0.040305. Since it is a rare event, the hypothesis H may not be true.

To discuss things without ambiguity, we use the following terms. Fix $0 < \alpha \ll 1$. If an event of probability less than α under the hypothesis H occurs, we say that H is rejected at the *significance level* α. If it does not occur, we say that H is accepted at the significance level α. The smaller the significance level α is, the more difficult to reject H. With these terms, possible answers to Exercise III are "the hypothesis H is rejected at the significance level 5%" and "the hypothesis H is accepted at the significance level 1%".

Let us present Exercise III in another form.

Exercise III′ In a certain region, 200 newborn babies were chosen at random, and 115 of them were boys. May we say that the ratio of boys and girls of newborn babies in this region is 1:1?

In the same way as above, one of the possible answers to Exercise III′ is that the hypothesis "the ratio of boys and girls of newborn babies in this region is 1:1" is rejected at the significance level 5%.

Chapter 4

Monte Carlo method

The history of the Monte Carlo method started when Ulam, von Neumann and others applied it to the simulation of nuclear fissions†[1] by a newly invented computer in 1940's, i.e., in the midst of World War II. Since then, along with the development of computer, the Monte Carlo method has been used in various fields of science and technology, and has produced remarkable results. The development of the Monte Carlo method will surely continue.

In this chapter, however, we do not mention such brilliant applications of the method, but we discuss its theoretical foundation. More concretely, through a solving process of Exercise I in Sec. 1.4, we study the basic theory and implementation of sampling of random variables by computer.

4.1 Monte Carlo method as gambling

It is of course desirable to solve mathematical problems by sure methods, but some problems such as extremely complicated ones or those that lack a lot of information can only be solved by stochastic methods. Those treated by the Monte Carlo method are such problems.

4.1.1 *Purpose*

The Monte Carlo method is a kind of *gambling* as its name indicates. The purpose of the player Alice, is to get a *generic value*—a typical value or not an exceptional value—of a given random variable by sampling. Of course, the mathematical problem in question is assumed to be solved by a generic value of the random variable. The sampling is done by her will, but she has

[1] In order to make the atomic bombs.

a risk to get an exceptional value, which risk should be measured in terms of probability.

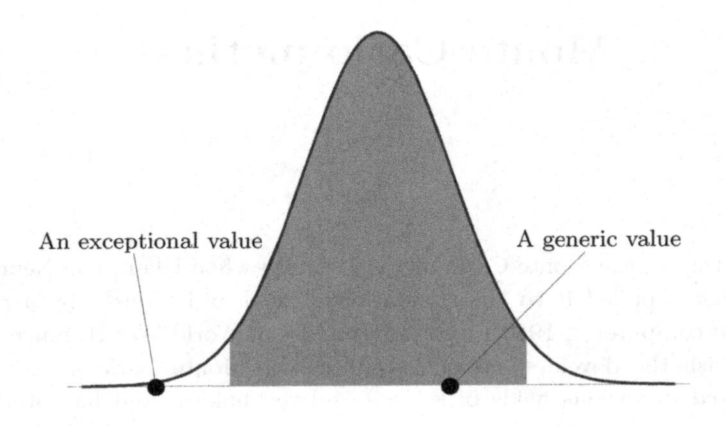

Fig. 4.1 The distribution and a generic value of a random variable (Conceptual figure)

The following is a very small example of the Monte Carlo method (without computer).

Example 4.1. An urn contains 100 balls, 99 of which are numbered r and one of which is numbered $r+1$. Alice draws a ball from the urn, and guesses the number r to be the number of her ball. The probability that she fails to guess the number r correctly is $1/100$.

If we state Example 4.1 in terms of gambling, it should be "Alice wins if she draws a generic ball, i.e., a ball numbered r, and loses otherwise. The probability that she loses is $1/100$."

In general, the player cannot know whether the aim has been attained or not even after the sampling. Indeed, in the above example, although the risk is measured, Alice cannot tell if her guess is correct even after drawing a ball.†[2]

[2]Many choices in our lives are certainly gambles. It is often the case that we do not know whether they were correct or not ....

4.1.2 *Exercise I, revisited*

To implement Example 4.1, we need only an urn and 100 numbered balls and nothing else, but actual Monte Carlo methods are implemented on such large scales that we need computers. Let us revisit Exercise I in Sec. 1.4.

> **Exercise I** When we toss a coin 100 times, what is the probability p that it comes up Heads at least 6 times in succession?

We apply the interval estimation in mathematical statistics. Repeat independent trials of "100 coin tosses" N times, and let S_N be the number of the occurrences of "the coin comes up Heads at least 6 times in succession" among the trials. Then, by the law of large numbers, the sample mean S_N/N is a good estimator for p when N is large. More concretely, we put $N := 10^6$.

Example 4.2. On the probability space $(\{0,1\}^{10^8}, \mathfrak{P}(\{0,1\}^{10^8}), P_{10^8})$, we construct the random variable S_{10^6} in the following way.

First, we define a function $X : \{0,1\}^{100} \to \{0,1\}$ by

$$X(\eta_1, \ldots, \eta_{100}) := \max_{1 \leqq r \leqq 100-5} \prod_{i=r}^{r+5} \eta_i, \qquad (\eta_1, \ldots, \eta_{100}) \in \{0,1\}^{100}.$$

This means that $X = 1$ if there are 6 successive 1's in $(\eta_1, \ldots, \eta_{100})$ and $X = 0$ otherwise. Next, we define $X_k : \{0,1\}^{10^8} \to \{0,1\}$, $k = 1, 2, \ldots, 10^6$, for $\omega = (\omega_1, \ldots, \omega_{10^8}) \in \{0,1\}^{10^8}$ by

$$\begin{aligned} X_1(\omega) &:= X(\omega_1, \ldots, \omega_{100}), \\ X_2(\omega) &:= X(\omega_{101}, \ldots, \omega_{200}), \\ &\vdots \end{aligned}$$

i.e., for each $k = 1, \ldots, 10^6$,

$$X_k(\omega) := X(\omega_{100(k-1)+1}, \ldots, \omega_{100k}).$$

$\{X_k\}_{k=1}^{10^6}$ are i.i.d. under P_{10^8}, and

$$P_{10^8}(X_k = 1) = p, \quad P_{10^8}(X_k = 0) = 1 - p,$$

i.e., $\{X_k\}_{k=1}^{10^6}$ are nothing but unfair coin tosses. Hence as in Example 3.4, we have

$$\mathbf{E}[X_k] = p, \quad \mathbf{V}[X_k] = p(1-p).$$

Finally, we define $S_{10^6} : \{0,1\}^{10^8} \to \mathbb{R}$ by

$$S_{10^6}(\omega) := \sum_{k=1}^{10^6} X_k(\omega), \quad \omega \in \{0,1\}^{10^8}.$$

Then, the mean and the variance of $S_{10^6}/10^6$ are, by (3.12) and (3.13),

$$\mathbf{E}\left[\frac{S_{10^6}}{10^6}\right] = p, \quad \mathbf{V}\left[\frac{S_{10^6}}{10^6}\right] = \frac{p(1-p)}{10^6} \leqq \frac{1}{4 \cdot 10^6}.$$

Let U_0 be a set of ω that give exceptional values to S_{10^6}:

$$U_0 := \left\{ \omega \in \{0,1\}^{10^8} \,\middle|\, \left| \frac{S_{10^6}(\omega)}{10^6} - p \right| \geqq \frac{1}{200} \right\}. \tag{4.1}$$

Now, Chebyshev's inequality shows

$$P_{10^8}(U_0) \leqq \frac{1}{4 \cdot 10^6} \cdot 200^2 = \frac{1}{100}. \tag{4.2}$$

Namely, a generic value of $S_{10^6}/10^6$ will be an approximate value of p.

Example 4.2 is regarded as gambling in the following way. Alice chooses an $\omega \in \{0,1\}^{10^8}$. If $\omega \notin U_0$, she wins and if $\omega \in U_0$, she loses. The probability that she loses is at most $1/100$.

The risk estimate (4.2) assumes that Alice is equally likely to choose $\omega \in \{0,1\}^{10^8}$. This assumption, however, can never be satisfied because she cannot choose ω from random numbers, which account for nearly all sequences. This is the most essential problem of sampling in the large-scale Monte Carlo method.

4.2 Pseudorandom generator

4.2.1 *Definition*

To play the gamble of Example 4.2, anyhow, Alice has to choose an $\omega \in \{0,1\}^{10^8}$. Let us suppose that she uses the most used device to do it, i.e., a pseudorandom generator.

Definition 4.1. A function $g : \{0,1\}^l \to \{0,1\}^n$ is called a *pseudorandom generator* if $l < n$. The input $\omega' \in \{0,1\}^l$ of g is called a *seed*, and the output $g(\omega') \in \{0,1\}^n$ a *pseudorandom number*.

To produce a pseudorandom number, we need to choose a seed $\omega' \in \{0,1\}^l$ of g, which procedure is called *initialization* (or *randomization*). For practical use, l should be so small that Alice can input any seed $\omega' \in \{0,1\}^l$

directly from a keyboard, and the program of the function g should work sufficiently fast.

Example 4.3. In Example 4.2, suppose that Alice uses a pseudorandom generator $g : \{0,1\}^{238} \to \{0,1\}^{10^8}$.†3 She chooses a seed $\omega' \in \{0,1\}^{238}$ of g and inputs it from a keyboard to a computer. Since ω' is only a 238 bit data ($\approx$ 30 letters of alphabet), it is easy to input from the keyboard. Then, the computer produces $S_{10^6}(g(\omega'))$.

The reason why Alice uses a pseudorandom generator is that $\omega \in \{0,1\}^{10^8}$ she has to choose is too long. If it is short enough, no pseudorandom generator is necessary. For example, when drawing a ball from the urn in Example 4.1, who on earth uses a pseudorandom generator?

4.2.2 *Security*

Let us continue to consider the case of Example 4.3. Alice can choose any seed $\omega' \in \{0,1\}^{238}$ of g freely of her own will. Her risk is now estimated by

$$P_{238}\left(\left|\frac{S_{10^6}(g(\omega'))}{10^6} - p\right| \geqq \frac{1}{200}\right). \tag{4.3}$$

Of course, the probability (4.3) depends on g. If this probability—the probability that her sample $S_{10^6}(g(\omega'))$ is an exceptional value of S—is large, then it is difficult for her to win the game, which is not desirable.

Now, we give the following (somewhat vague) definition: we say a pseudorandom generator $g : \{0,1\}^l \to \{0,1\}^n$, $l < n$, is *secure against a set* $U \subset \{0,1\}^n$ if it holds that

$$P_n(\omega \in U) \approx P_l(g(\omega') \in U).$$

In Example 4.3, if $g : \{0,1\}^{238} \to \{0,1\}^{10^8}$ is secure against U_0 in (4.1), for the majority of the seeds $\omega' \in \{0,1\}^{238}$ that Alice can choose of her own will, the samples $S_{10^6}(g(\omega'))$ will be generic values of S_{10^6}. In this case, no random number is needed. In other words, in sampling a value of S_{10^6}, using g does not make Alice's risk high, and hence g is said to be secure. The problem of sampling in a Monte Carlo method will be solved by finding a suitable secure pseudorandom generator.

In general, a pseudorandom generator $g : \{0,1\}^l \to \{0,1\}^n$ is desired to be secure against as many subsets of $\{0,1\}^n$ as possible, but there is

[3] The origin of the number 238 will soon be clear in Example 4.4.

no pseudorandom generator that is secure against all subsets of $\{0,1\}^n$. Indeed, for any $g:\{0,1\}^l \to \{0,1\}^n$, let

$$U_g := \{\, g(\omega') \mid \omega' \in \{0,1\}^l \,\} \subset \{0,1\}^n.$$

Then, we have $P_n(\omega \in U_g) \leq 2^{l-n}$ but $P_l(g(\omega') \in U_g) = 1$, which means that g is not secure against U_g.

4.3 Monte Carlo integration

For a while, leaving Exercise I aside, let us consider a general problem. Let X be a function of m coin tosses, i.e., $X:\{0,1\}^m \to \mathbb{R}$, and consider calculating the mean

$$\mathbf{E}[X] = \frac{1}{2^m} \sum_{\eta \in \{0,1\}^m} X(\eta)$$

of X numerically. When m is small, we can directly calculate the finite sum of the right-hand side, but when m is large, e.g., $m = 100$, the direct calculation becomes impossible in practice because of the huge amount of computation. In such a case, we estimate the mean of X applying the law of large numbers, which is called the *Monte Carlo integration* (Example 4.2). Most of scientific Monte Carlo methods aim at calculating some characteristics of distributions of random variables, e.g., means, variances, etc., which are actually Monte Carlo integrations.

4.3.1 *Mean and integral*

In general, mean is considered to be integral. Indeed, for a function X of m coin tosses, we define

$$\hat{X}(x) := X(d_1(x), \ldots, d_m(x)), \quad x \in [0,1), \tag{4.4}$$

where $\{d_i(x)\}_{i=1}^m$ are Borel's model of m coin tosses (Example 1.2). Then, we have

$$\mathbf{E}[X] = \int_0^1 \hat{X}(x)dx. \tag{4.5}$$

The Monte Carlo 'integration' is named after this fact.

Let us show (4.5). Note that $\hat{X}:[0,1) \to \mathbb{R}$ is a step function (Fig. 4.2):

$$\hat{X}(x) = \sum_{i=0}^{2^m-1} X(d_1(2^{-m}i), \ldots, d_m(2^{-m}i))\mathbf{1}_{[2^{-m}i,\, 2^{-m}(i+1))}(x), \quad x \in [0,1).$$

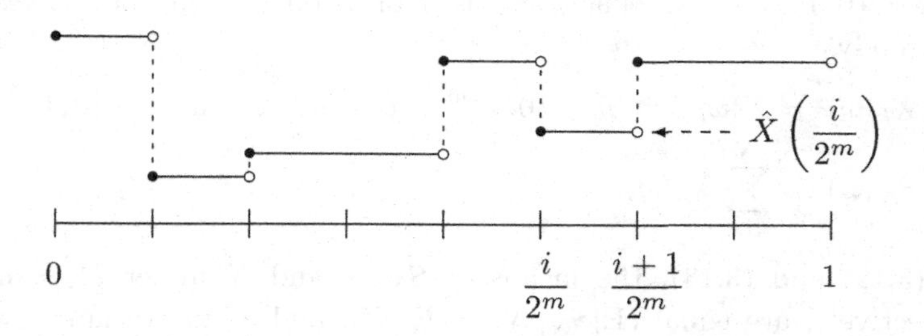

Fig. 4.2 The graph of $\hat{X}(x)$ (Example)

Here $\mathbf{1}_{[2^{-m}i,\,2^{-m}(i+1))}(x)$ is the indicator function of the interval $[2^{-m}i,\,2^{-m}(i+1))$.

Then, we have

$$\begin{aligned}\int_0^1 \hat{X}(x)dx &= \sum_{i=0}^{2^m-1} X(d_1(2^{-m}i),\ldots,d_m(2^{-m}i))\int_0^1 \mathbf{1}_{[2^{-m}i,\,2^{-m}(i+1))}(x)dx\\ &= \sum_{i=0}^{2^m-1} X(d_1(2^{-m}i),\ldots,d_m(2^{-m}i))\cdot 2^{-m}\\ &= \frac{1}{2^m}\sum_{\eta\in\{0,1\}^m} X(\eta) \;=\; \mathbf{E}[X].\end{aligned}$$

Lemma 4.1. *Let $X : \{0,1\}^m \to \mathbb{R}$ and let $\hat{X} : [0,1) \to \mathbb{R}$ be the corresponding function defined by (4.4). Then, for any $j \in \mathbb{N}_+$, we have*

$$\int_0^1 \hat{X}(x)dx \;=\; \frac{1}{2^{m+j}}\sum_{q=0}^{2^{m+j}-1} \hat{X}\left(\frac{q}{2^{m+j}}\right).$$

Proof. It is enough to prove the lemma for $\hat{X}(x) = \mathbf{1}_{[2^{-m}i,\,2^{-m}(i+1))}(x)$.

$$\begin{aligned}\frac{1}{2^{m+j}}\sum_{q=0}^{2^{m+j}-1} \mathbf{1}_{[2^{-m}i,\,2^{-m}(i+1))}\left(\frac{q}{2^{m+j}}\right) &= \frac{1}{2^{m+j}}\sum_{q=2^j i}^{2^j(i+1)-1} 1\\ &= \frac{1}{2^m}.\end{aligned}$$

□

4.3.2 *Estimation of mean*

We state Example 4.2 in a more general setup. Let S_N be the sum of i.i.d. random variables $\{X_k\}_{k=1}^N$ whose common distribution is identical to that of $X : \{0,1\}^m \to \mathbb{R}$. Then, S_N is a function of Nm coin tosses, more concretely,

$$X_k(\omega) := X(\omega_k), \quad \omega_k \in \{0,1\}^m, \quad \omega = (\omega_1, \ldots, \omega_N) \in \{0,1\}^{Nm},$$
$$S_N(\omega) := \sum_{k=1}^{N} X_k(\omega). \tag{4.6}$$

By (3.12) and (3.13), the means of S_N/N and X under P_{Nm} and P_m, respectively, are equal ($\mathbf{E}[S_N/N] = \mathbf{E}[X]$), and as for variance, we have $\mathbf{V}[S_N/N] = \mathbf{V}[X]/N$.

For $N \gg 1$, a generic value of S_N/N will be an approximate value of $\mathbf{E}[X]$ by the law of large numbers. Let us estimate the risk by Chebyshev's inequality:

$$P_{Nm}\left(\left| \frac{S_N(\omega)}{N} - \mathbf{E}[X] \right| \geqq \delta \right) \leqq \frac{\mathbf{V}[X]}{N\delta^2}. \tag{4.7}$$

This means that putting

$$U_1 := \left\{ \omega \in \{0,1\}^{Nm} \;\middle|\; \left| \frac{S_N(\omega)}{N} - \mathbf{E}[X] \right| \geqq \delta \right\}, \tag{4.8}$$

we are considering a gamble "if Alice's choice $\omega \in \{0,1\}^{Nm}$ is not in U_1, she wins, if $\omega \in U_1$, she loses".

In general, as $\mathbf{E}[X]$ is unknown, $\mathbf{V}[X]$ is also unknown. In this sense, the risk estimate (4.7) is not complete. However, if we can find a constant $M > 0$ such that $\mathbf{V}[X] \leqq M$, the risk estimate then becomes complete (Example 4.2).

Of course, the Monte Carlo integration can be applied to not only random variables but also functions that have no relation to probability. If a given integrand is not so complicated as X in Example 4.2, then deterministic sampling methods, such as the quasi-Monte Carlo method, may be applicable. See [Bouleau and Lépingle (1994)] for details.

4.3.3 *Random Weyl sampling*

In the Monte Carlo integration, the random variable S_N in (4.6) has a special form. Using this fact, we can construct a pseudorandom generator that is secure against U_1 in (4.8).

First, we introduce some notations. For each $m \in \mathbb{N}_+$, we define

$$D_m := \{ 2^{-m} i \mid i = 0, \ldots, 2^m - 1 \} \subset [0,1).$$

Let $P_{(m)}$ denote the uniform probability measure on D_m. For each $x \geqq 0$, we define

$$\lfloor x \rfloor_m := \lfloor 2^m (x - \lfloor x \rfloor) \rfloor \cdot 2^{-m} \in D_m.$$

Namely, $\lfloor x \rfloor_m$ denotes the truncation of x to m decimal place in its binary expansion. By a one-to-one correspondence

$$D_m \ni 2^{-m} i \longleftrightarrow (d_1(2^{-m} i), \ldots, d_m(2^{-m} i)) \in \{0,1\}^m,$$

we identify D_m with $\{0,1\}^m$, and write it as $D_m \cong \{0,1\}^m$.

Definition 4.2. Let $j \in \mathbb{N}_+$. We define

$$\begin{aligned} Z_k(\omega') &:= \lfloor x + k\alpha \rfloor_m \in D_m, \\ \omega' &= (x, \alpha) \in D_{m+j} \times D_{m+j} \cong \{0,1\}^{2m+2j}, \\ & \qquad k = 1, 2, 3, \ldots, 2^{j+1}. \end{aligned}$$

Then, for $N \leqq 2^{j+1}$, we define a pseudorandom generator

$$g : \{0,1\}^{2m+2j} \to \{0,1\}^{Nm} \tag{4.9}$$

by

$$g(\omega') := (Z_1(\omega'), \ldots, Z_N(\omega')) \in {D_m}^N \cong \{0,1\}^{Nm}.$$

Lemma 4.2. *Under the product (uniform) probability measure $P_{(m+j)} \otimes P_{(m+j)}$ on $D_{m+j} \times D_{m+j}$, the random variables $\{Z_k\}_{k=1}^{2^{j+1}}$ are pairwise independent (Remark 3.4), and each Z_k is uniformly distributed in D_m.*

Proof. Take arbitrary $1 \leqq k < k' \leqq 2^{j+1}$ and arbitrary $t,\ t' \in D_m$. Then, it is enough to show that

$$P_{(m+j)} \otimes P_{(m+j)} \left(Z_k = t,\ Z_{k'} = t' \right) = \frac{1}{2^{2m}}. \tag{4.10}$$

Indeed, (4.10) implies

$$\begin{aligned} P_{(m+j)} \otimes P_{(m+j)} \left(Z_k = t \right) &= \sum_{t' \in D_m} P_{(m+j)} \otimes P_{(m+j)} \left(Z_k = t,\ Z_{k'} = t' \right) \\ &= 2^m \cdot \frac{1}{2^{2m}} = \frac{1}{2^m}, \end{aligned}$$

and hence each Z_k is uniformly distributed in D_m. Then, since

$$\begin{aligned} & P_{(m+j)} \otimes P_{(m+j)} \left(Z_k = t,\ Z_{k'} = t' \right) \\ & \qquad = P_{(m+j)} \otimes P_{(m+j)} \left(Z_k = t \right) \times P_{(m+j)} \otimes P_{(m+j)} \left(Z_{k'} = t' \right) \end{aligned}$$

Z_k and $Z_{k'}$ are independent.

We rewrite (4.10) in an integral form: defining two periodic functions $F, G : [0, \infty) \to \{0, 1\}$ with period 1 by

$$F(x) := \mathbf{1}_{[t', t'+2^{-m})}(x - \lfloor x \rfloor),$$
$$G(x) := \mathbf{1}_{[t, t+2^{-m})}(x - \lfloor x \rfloor),$$

we will show the following equality, which is equivalent to (4.10).

$$\mathbf{E}[F(Z_{k'})G(Z_k)] = \int_0^1 F(u)du \int_0^1 G(v)dv. \tag{4.11}$$

Here $\mathbf{E}$ stands for the mean under $P_{(m+j)} \otimes P_{(m+j)}$.

Since $Z_{k'}$ and Z_k are random variables on $(D_{m+j} \times D_{m+j}, \mathfrak{P}(D_{m+j} \times D_{m+j}), P_{(m+j)} \otimes P_{(m+j)})$, the mean of $F(Z_{k'})G(Z_k)$ is calculated as follows.

$$\begin{aligned}
&\mathbf{E}[F(Z_{k'})G(Z_k)] \\
&= \frac{1}{2^{m+j}} \sum_{\alpha \in D_{m+j}} \frac{1}{2^{m+j}} \sum_{x \in D_{m+j}} F(\lfloor x + k'\alpha \rfloor_m) G(\lfloor x + k\alpha \rfloor_m) \\
&= \frac{1}{2^{m+j}} \sum_{\alpha \in D_{m+j}} \frac{1}{2^{m+j}} \sum_{x \in D_{m+j}} F(x + k'\alpha) G(x + k\alpha) \\
&= \frac{1}{2^{2m+2j}} \sum_{q=0}^{2^{m+j}-1} \sum_{p=0}^{2^{m+j}-1} F\left(\frac{p}{2^{m+j}} + \frac{k'q}{2^{m+j}}\right) G\left(\frac{p}{2^{m+j}} + \frac{kq}{2^{m+j}}\right) \\
&= \frac{1}{2^{2m+2j}} \sum_{q=0}^{2^{m+j}-1} \sum_{p=kq}^{2^{m+j}+kq-1} F\left(\frac{p}{2^{m+j}} + \frac{(k'-k)q}{2^{m+j}}\right) G\left(\frac{p}{2^{m+j}}\right) \\
&= \frac{1}{2^{2m+2j}} \sum_{q=0}^{2^{m+j}-1} \sum_{p=0}^{2^{m+j}-1} F\left(\frac{p}{2^{m+j}} + \frac{(k'-k)q}{2^{m+j}}\right) G\left(\frac{p}{2^{m+j}}\right).
\end{aligned} \tag{4.12}$$

We here assume that $0 < k' - k = 2^i s \leqq 2^{j+1} - 1$, where $0 \leqq i \leqq j$ and s is an odd integer. Then, we see

$$\frac{1}{2^{m+j}} \sum_{q=0}^{2^{m+j}-1} F\left(\frac{p}{2^{m+j}} + \frac{(k'-k)q}{2^{m+j}}\right) = \frac{1}{2^{m+j}} \sum_{q=0}^{2^{m+j}-1} F\left(\frac{p}{2^{m+j}} + \frac{sq}{2^{m+j-i}}\right). \tag{4.13}$$

Now, we need an algebraic argument. For $q, q' \in \{0, 1, 2, \dots, 2^{m+j-i} - 1\}$, if we have

$$sq \bmod 2^{m+j-i} = sq' \bmod 2^{m+j-i} \quad \dagger 4$$

[4] $a \bmod m$ stands for the remainder on division of a by m.

then

$$s\,(q-q') \bmod 2^{m+j-i} = 0,$$

i.e., $s\,(q-q')$ is divisible by 2^{m+j-i}. Since s is odd, $q-q'$ is divisible by 2^{m+j-i}, but $q,\,q' \in \{0,1,2,\ldots,2^{m+j-i}-1\}$, it means $q=q'$. Therefore a correspondence

$$\{0,1,2,\ldots,2^{m+j-i}-1\} \ni q \longleftrightarrow sq \bmod 2^{m+j-i} \in \{0,1,2,\ldots,2^{m+j-i}-1\}$$

is one-to-one. Let $q_r \in \{0,1,2,\ldots,2^{m+j-i}-1\}$ be a unique solution to

$$sq_r \bmod 2^{m+j-i} = r, \quad r \in \{0,1,2,\ldots,2^{m+j-i}-1\}.$$

Then, for each r, we have

$$\begin{aligned}
&\#\{0 \leqq q \leqq 2^{m+j}-1 \mid sq \bmod 2^{m+j-i} = r\}\\
&\quad = \#\{0 \leqq q \leqq 2^{m+j}-1 \mid sq \bmod 2^{m+j-i} = sq_r \bmod 2^{m+j-i}\}\\
&\quad = \#\{0 \leqq q \leqq 2^{m+j}-1 \mid q-q_r \text{ is divisible by } 2^{m+j-i}\} = 2^i.
\end{aligned}$$

From this, (4.13) is calculated as

$$\begin{aligned}
\frac{1}{2^{m+j}}\sum_{q=0}^{2^{m+j}-1} F\left(\frac{p}{2^{m+j}}+\frac{sq}{2^{m+j-i}}\right) &= \frac{2^i}{2^{m+j}}\sum_{r=0}^{2^{m+j-i}-1} F\left(\frac{p}{2^{m+j}}+\frac{r}{2^{m+j-i}}\right)\\
&= \frac{1}{2^{m+j-i}}\sum_{r=0}^{2^{m+j-i}-1} F\left(\frac{r}{2^{m+j-i}}\right)\\
&= \int_0^1 F(u)du. \qquad (4.14)
\end{aligned}$$

The last '=' is due to Lemma 4.1. From (4.12), (4.13), and (4.14), it follows that

$$\begin{aligned}
&\mathbf{E}[F(Z_{k'})G(Z_k)]\\
&= \frac{1}{2^{m+j}}\sum_{p=0}^{2^{m+j}-1}\left(\frac{1}{2^{m+j}}\sum_{q=0}^{2^{m+j}-1} F\left(\frac{p}{2^{m+j}}+\frac{(k'-k)q}{2^{m+j}}\right)\right) G\left(\frac{p}{2^{m+j}}\right)\\
&= \frac{1}{2^{m+j}}\sum_{p=0}^{2^{m+j}-1}\left(\frac{1}{2^{m+j}}\sum_{q=0}^{2^{m+j}-1} F\left(\frac{p}{2^{m+j}}+\frac{sq}{2^{m+j-i}}\right)\right) G\left(\frac{p}{2^{m+j}}\right)\\
&= \int_0^1 F(u)du\cdot\frac{1}{2^{m+j}}\sum_{p=0}^{2^{m+j}-1} G\left(\frac{p}{2^{m+j}}\right)\\
&= \int_0^1 F(u)du\int_0^1 G(v)dv.
\end{aligned}$$

This completes the proof of (4.11). □

Theorem 4.1. *The pseudorandom generator* $g : \{0,1\}^{2m+2j} \to \{0,1\}^{Nm}$ *in (4.9) satisfies that for* S_N *in (4.6),*

$$\mathbf{E}[S_N(g(\omega'))] = \mathbf{E}[S_N(\omega)]\ (= N\mathbf{E}[X]),$$
$$\mathbf{V}[S_N(g(\omega'))] = \mathbf{V}[S_N(\omega)]\ (= N\mathbf{V}[X]).$$

Here ω' *and* ω *are assumed to be uniformly distributed in* $\{0,1\}^{2m+2j}$ *and in* $\{0,1\}^{Nm}$*, respectively. From this, just like (4.7), Chebyshev's inequality*

$$P_{2m+2j}(g(\omega') \in U_1) = P_{2m+2j}\left(\left| \frac{S_N(g(\omega'))}{N} - \mathbf{E}[X] \right| \geq \delta \right) \leq \frac{\mathbf{V}[X]}{N\delta^2}$$

follows. In this sense, g *is secure against* U_1 *in (4.8).*

The Monte Carlo integration method using the secure pseudorandom generator (4.9) is called the *random Weyl sampling.*†[5]

Proof of Theorem 4.1. First, since each $Z_k(\omega')$ is uniformly distributed in $\{0,1\}^m$, we see

$$\mathbf{E}[S_N(g(\omega'))] = \mathbf{E}\left[\sum_{k=1}^{N} X(Z_k(\omega'))\right] = N\mathbf{E}[X].$$

Next, as we mentioned in Remark 3.4, the pairwise independence of $\{Z_k\}_{k=1}^N$ implies that

$$\begin{aligned}
\mathbf{V}[S_N(g(\omega'))] &= \mathbf{E}\left[\left(\sum_{k=1}^{N}(X(Z_k(\omega')) - \mathbf{E}[X])\right)^2\right] \\
&= \sum_{k=1}^{N}\sum_{k'=1}^{N} \mathbf{E}\left[(X(Z_k(\omega')) - \mathbf{E}[X])(X(Z_{k'}(\omega')) - \mathbf{E}[X])\right] \\
&= \sum_{k=1}^{N} \mathbf{E}\left[(X(Z_k(\omega')) - \mathbf{E}[X])^2\right] \\
&\quad +2 \sum_{1 \leq k < k' \leq N} \mathbf{E}\left[(X(Z_k(\omega')) - \mathbf{E}[X])(X(Z_{k'}(\omega')) - \mathbf{E}[X])\right] \\
&= \sum_{k=1}^{N} \mathbf{E}\left[(X(Z_k(\omega')) - \mathbf{E}[X])^2\right] \\
&= N\mathbf{V}[X].
\end{aligned}$$

Thus we know that g has the required properties. □

[5] When $0 < \alpha < 1$ is an irrational number, the transformation $[0,1) \ni x \mapsto x + \alpha - \lfloor x + \alpha \rfloor \in [0,1)$ is called the *Weyl transformation*, after which this pseudorandom generator is named.

Example 4.4. Applying the random Weyl sampling, we can solve Exercise I. Let S_{10^6} be the random variable defined in Example 4.2. We take $m := 100$ and $N := 10^6$ in Definition 4.2. In order to let $N \leqq 2^{j+1}$, it is enough to take $j := 19$. Then, we have $2m + 2j = 238$ so that the pseudorandom generator (4.9) is now a function $g : \{0,1\}^{238} \to \{0,1\}^{10^8}$. The risk (4.3) is estimated, by Theorem 4.1, as

$$P_{238}\left(\left| \frac{S_{10^6}(g(\omega'))}{10^6} - p \right| \geqq \frac{1}{200} \right) \leqq \frac{1}{100}. \tag{4.15}$$

Thus g is secure against U_0 in (4.1). Since Alice can freely choose any seed $\omega' \in \{0,1\}^{238}$, this risk estimate has a practical meaning and she no longer needs a long random number.

Here is a concrete example. Instead of her, the author chose the following seed $\omega' = (x, \alpha) \in D_{119} \times D_{119} \cong \{0,1\}^{238}$ written in the binary numeral system:

$$\begin{aligned} x &= 0.1110110101\ 1011101101\ 0100000011\ 0110101001\ 0101000100 \\ &\quad\ 0101111101\ 1010000000\ 1010100011\ 0100011001\ 1101111101 \\ &\quad\ 1101010011\ 111100100, \end{aligned}$$

$$\begin{aligned} \alpha &= 0.1100000111\ 0111000100\ 0001101011\ 1001000001\ 0010001000 \\ &\quad\ 1010101101\ 1110101110\ 0010010011\ 1000000011\ 0101000110 \\ &\quad\ 0101110010\ 010111111. \end{aligned}$$

Then, we obtained $S_{10^6}(g(\omega')) = 546,177$ by a computer (see Sec. A.5). In this case,

$$\frac{S_{10^6}(g(\omega'))}{10^6} = 0.546177 \tag{4.16}$$

is the estimated value of the probability p. This result with the risk estimate (4.15) is expressed in statistical terms as "The 99% confidence interval of p is 0.546177 ± 0.05, i.e., $0.541 < p < 0.551$". Indeed, the true value of p is

$$692255904222999797557597756576 \times 2^{-100} = 0.5460936192.$$

Thus the error of the estimated value (4.16) is only 0.00008.

In a practical Monte Carlo integration, usually, the sample size N is not determined in advance, but it is determined in doing numerical experiments. To be ready for such a situation, we should take j somewhat large.

Remark 4.1. We can advise Alice a little in choosing a seed $\omega' = (x, \alpha) \in \{0,1\}^{2m+2j}$ for the random Weyl sampling: she should not choose an extremely simple α. Indeed, if she chooses $\alpha = (0,0,\ldots,0) \in \{0,1\}^{m+j}$, the sampling will certainly end in failure.

4.4 From the viewpoint of mathematical statistics

We have formulated the Monte Carlo method as gambling, and we have assumed that Alice chooses a seed $\omega' \in \{0,1\}^l$ of a pseudorandom generator $g : \{0,1\}^l \to \{0,1\}^n$ of her own will. From the viewpoint of mathematical statistics, this is not a good formulation because sampling should be done randomly in order to guarantee the objectivity of the result. Indeed, in the case of the random Weyl sampling, as we mentioned in Remark 4.1, Alice can choose a bad seed on purpose, i.e., the result may depend on the player's will.

Of course, it is impossible to discuss the objectivity of sampling rigorously. We here simply assume that Heads or Tails of coin tosses do not depend on anyone's will (Remark 1.3). Then, for example, when we choose a seed $\omega' \in \{0,1\}^l$, we toss a coin l times, record 1 if it comes up Heads and 0 if it comes up Tails at each coin toss, and define ω' as the recorded $\{0,1\}$-sequence, which completes an objective sampling. As a matter of fact, the seed $\omega' \in \{0,1\}^{238}$ in Example 4.4 was chosen in this way.

This method cannot be used to choose a very long $\omega \in \{0,1\}^n$. The point is that keeping the risk low, the random Weyl sampling makes the input ω' much shorter so that this method may become executable.

Appendix A

A.1 Symbols and terms

A.1.1 *Set and function*

Definition A.1. For two sets A and B, the set of all ordered pairs (x, y) of $x \in A$ and $y \in B$ is written as

$$A \times B := \{(x, y) \mid x \in A,\ y \in B\},$$

and it is called the *direct product* of A and B.

Example A.1. For two intervals $A = [a, b] := \{\, x \mid a \leqq x \leqq b \,\}$ and $B = [c, d]$ of real line, their direct product $A \times B$ is a rectangle in the coordinate plane.

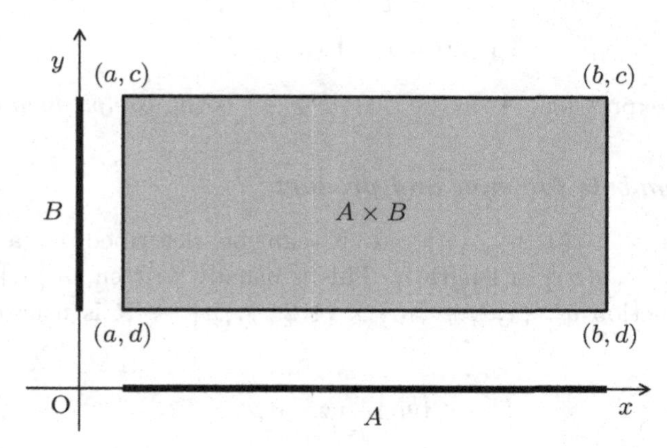

For more than two sets, the direct product is similarly defined. For example,

$$\mathbb{R}^3 = \mathbb{R} \times \mathbb{R} \times \mathbb{R} := \{(x, y, z) \mid x \in \mathbb{R},\ y \in \mathbb{R},\ z \in \mathbb{R}\}$$

is the set of all ordered triplets of real numbers, i.e., the set of all points in 3-dimensional space. $\{0, 1\}^3$ in Example 1.1 is nothing but $\{0, 1\} \times \{0, 1\} \times \{0, 1\}$.

Definition A.2. Let E and F be non-empty sets. If for each element of E, there corresponds an element of F, we call the correspondence a *function* (or a *mapping*), and write it as $f : E \to F$. When $E = F$, we also call it a *transformation*. The corresponding element of F to an individual $a \in E$ is written as $f(a)$, and this individual correspondence is written as $a \mapsto f(a)$.

Let $\Omega \neq \emptyset$ be a set.[†6] For $A \in \mathfrak{P}(\Omega)$, define a function $\mathbf{1}_A : \Omega \to \{0,1\}$ by

$$\mathbf{1}_A(\omega) = \begin{cases} 1 \ (\omega \in A), \\ 0 \ (\omega \notin A), \end{cases}$$

which is called the *indicator function* of A. Relations and operations for sets are realized as relations and calculations of indicator functions. For example, "$A \subset B$" is equivalent to "$\mathbf{1}_A(\omega) \leqq \mathbf{1}_B(\omega),\ \omega \in \Omega$". Among others,

$$\begin{aligned} \mathbf{1}_{A\cup B}(\omega) &= \max\{\mathbf{1}_A(\omega), \mathbf{1}_B(\omega)\}, \\ \mathbf{1}_{A\cap B}(\omega) &= \min\{\mathbf{1}_A(\omega), \mathbf{1}_B(\omega)\}, \\ &= \mathbf{1}_A(\omega) \cdot \mathbf{1}_B(\omega), \\ \mathbf{1}_{A^c}(\omega) &= 1 - \mathbf{1}_A(\omega). \end{aligned}$$

In the last expression, $A^c := \{\omega \in \Omega \,|\, \omega \notin A\}$ is the *complement* of A.

A.1.2 *Symbols for sum and product*

A function $a : \{1, 2, \ldots, n\} \to \mathbb{R}$ can be described as a sequence $\{a(1), a(2), \ldots, a(n)\}$ of length n. This is usually written as $\{a_i\}_{i=1}^n$. Similarly, a function $b : \{1, 2, \ldots, m\} \times \{1, 2, \ldots, n\} \to \mathbb{R}$ is usually written as

$$\{b_{ij}\}_{\substack{i=1,2,\ldots,m, \\ j=1,2,\ldots,n}}.$$

This is called a double sequence. As we write the sum of a sequence $\{a_i\}_{i=1}^n$ as $\sum_{i=1}^n a_i$, the sum of a double sequence $\{b_{ij}\}_{\substack{i=1,2,\ldots,m, \\ j=1,2,\ldots,n}}$ is written as

$$\sum_{\substack{i=1,2,\ldots,m, \\ j=1,2,\ldots,n}} b_{ij}.$$

[6] '$\emptyset$' denotes the empty set.

This is called a double sum. Obviously,

$$\sum_{\substack{i=1,2,\ldots,m,\\ j=1,2,\ldots,n}} b_{ij} = \sum_{i=1}^{m}\left(\sum_{j=1}^{n} b_{ij}\right) = \sum_{j=1}^{n}\left(\sum_{i=1}^{m} b_{ij}\right).$$

Similarly, triple sequence and triple sum (more generally, multiple sequence and multiple sum) are defined.

The product $a_1 \times a_2 \times \cdots \times a_n$ of a sequence $\{a_i\}_{i=1}^n$ is written as

$$\prod_{i=1}^{n} a_i.$$

Similarly, the product of a double sequence $\{b_{ij}\}_{i=1,2,\ldots,m,\, j=1,2,\ldots,n}$ is written as

$$\prod_{\substack{i=1,2,\ldots,m,\\ j=1,2,\ldots,n}} b_{ij}.$$

Obviously,

$$\prod_{\substack{i=1,2,\ldots,m,\\ j=1,2,\ldots,n}} b_{ij} = \prod_{i=1}^{m}\left(\prod_{j=1}^{n} b_{ij}\right) = \prod_{j=1}^{n}\left(\prod_{i=1}^{m} b_{ij}\right).$$

Example A.2. Let us look at a little complicated double sequence. Suppose that for each $i = 1, \ldots, m$, a sequence $\{a_{ij}\}_{j=1}^{n_i}$ of length n_i is given. Then, the product

$$(a_{11}+a_{12}+\cdots+a_{1n_1})(a_{21}+a_{22}+\cdots+a_{2n_2})\times\cdots\times(a_{m1}+a_{m2}+\cdots+a_{mn_m})$$

can be written in terms of $\sum$ and $\prod$ as

$$\prod_{i=1}^{m}\sum_{j=1}^{n_i} a_{ij}. \tag{A.1}$$

Developing it, we obtain

$$\sum_{j_1=1}^{n_1}\cdots\sum_{j_m=1}^{n_m}\prod_{i=1}^{m} a_{ij_i}. \tag{A.2}$$

Conversely, resolving (A.2) into factors, we obtain (A.1).

The summation symbol $\sum$ is applied not only to sequences but also to any finite set of numbers. For example, suppose that for each element ω of a finite set Ω there corresponds a number $p_\omega \in \mathbb{R}$. Then, the total sum of such p_ω is written as

$$\sum_{\omega \in \Omega} p_\omega.$$

The sum of p_ω over $\omega \in \Omega$ that satisfies a condition $X(\omega) = a_i$ is written as

$$\sum_{\omega \in \Omega\,;\, X(\omega)=a_i} p_\omega.$$

The product symbol $\prod$ is used similarly.

A.1.3 *Inequality symbol '$\gg$'*

$a \gg b$ as well as $b \ll a$ means that a is much greater than b. Without making it clear how much a is greater than b, we use the symbol '$\gg$' to express "a is much greater than b". It may not sound mathematical, but if you look at how it is used you will accept it: "Since $\lim_{x\to\infty} x^{100}/e^x = 0$ (Proposition A.3 (i)), we have $x^{100} \ll e^x$ for $x \gg 1$."

A.2 Binary numeral system

To describe numbers, we usually use the decimal numeral system (or the base-10 numeral system). It is a positional numeral system employing 10 as the base and requiring 10 different numerals, the digits 0, 1, 2, 3, 4, 5, 6, 7, 8, 9. It also requires a dot (decimal point) to represent decimal fractions. The same thing can be done with only two different numerals , the digits 0, 1. This is called the binary numeral system (or the base-2 numeral system). It is Leibniz who first systematized it in mathematics.

The binary numeral system is the simplest positional numeral system, which can be expressed by 'ON'(= 1) and 'OFF'(= 0) of electronic circuits, so that it is now a mathematical basis of all digital technologies.

A.2.1 *Binary integers*

Let $D_i^{(10)}(n) \in \{0, 1, \ldots, 9\}$ denote the i-th digit of $n \in \mathbb{N}$ in the decimal numeral system. Then, we have

$$n = \sum_{i=1}^{\infty} 10^{i-1} D_i^{(10)}(n),$$

which is actually a finite sum for each n. For example,

$$563 = 10^2 \times 5 + 10^1 \times 6 + 10^0 \times 3.$$

Note that

$$D_i^{(10)}(n) := \lfloor 10^{-i+1} n \rfloor - 10 \lfloor 10^{-i} n \rfloor, \qquad i \in \mathbb{N}_+,$$

where $\lfloor t \rfloor$ denotes the integer part of $t \geqq 0$. For example,

$$\begin{aligned} D_2^{(10)}(563) &= \lfloor 10^{-1} \times 563 \rfloor - 10 \times \lfloor 10^{-2} \times 563 \rfloor \\ &= \lfloor 56.3 \rfloor - 10 \times \lfloor 5.63 \rfloor = 56 - 50 \\ &= 6. \end{aligned}$$

Similarly, for each $n \in \mathbb{N}$, we can find $D_i(n) \in \{0, 1\}$, $i \in \mathbb{N}$, so that

$$n = \sum_{i=1}^{\infty} 2^{i-1} D_i(n),$$

which is actually a finite sum. Indeed, they are given by

$$D_i(n) := \lfloor 2^{-i+1} n \rfloor - 2 \lfloor 2^{-i} n \rfloor, \qquad i \in \mathbb{N}.$$

Example A.3. For example,

$$\begin{aligned} D_1(563) &= \lfloor 563 \rfloor - 2 \times \lfloor 2^{-1} \times 563 \rfloor = 563 - 562 \\ &= 1, \\ D_2(563) &= \lfloor 2^{-1} \times 563 \rfloor - 2 \times \lfloor 2^{-2} \times 563 \rfloor \\ &= \lfloor 281.5 \rfloor - 2 \times \lfloor 140.75 \rfloor = 281 - 280 \\ &= 1. \end{aligned}$$

This computation needs some patience. There is an easier method to obtain binary integers from decimal integers: repeating the division by 2, and line the remainders up in the reverse order.

$$\begin{array}{rll} 2\,) & \underline{563} & \\ 2\,) & \underline{281} & \ldots 1 \\ 2\,) & \underline{140} & \ldots 1 \\ 2\,) & \underline{70} & \ldots 0 \\ 2\,) & \underline{35} & \ldots 0 \\ 2\,) & \underline{17} & \ldots 1 \\ 2\,) & \underline{8} & \ldots 1 \\ 2\,) & \underline{4} & \ldots 0 \\ 2\,) & \underline{2} & \ldots 0 \\ 2\,) & \underline{1} & \ldots 0 \\ 2\,) & \underline{0} & \ldots 1 \end{array}$$

Thus the binary representation of 563 is 1000110011.

A.2.2 *Binary fractions*

Next, let us consider how to describe fraction $x \in [0, 1)$. In the decimal numeral system,

$$x = 0.d_1^{(10)}(x)d_2^{(10)}(x)\ldots = \sum_{i=1}^{\infty} 10^{-i} d_i^{(10)}(x).$$

Here $d_i^{(10)}(x) \in \{0, 1, \ldots, 9\}$ is the i-th digit of x in its decimal expansion. For example,

$$0.563 = 10^{-1} \times 5 + 10^{-2} \times 6 + 10^{-3} \times 3 + 10^{-4} \times 0 + 10^{-5} \times 0 + \cdots.$$

$d_i^{(10)}(x)$ is described as

$$d_i^{(10)}(x) := \lfloor 10^i x \rfloor - 10 \lfloor 10^{i-1} x \rfloor, \qquad i \in \mathbb{N}. \tag{A.3}$$

For example,

$$\begin{aligned} d_2^{(10)}(0.563) &= \lfloor 100 \times 0.563 \rfloor - 10 \times \lfloor 10 \times 0.563 \rfloor \\ &= \lfloor 56.3 \rfloor - 10 \times \lfloor 5.63 \rfloor \\ &= 56 - 10 \times 5 = 6. \end{aligned}$$

In the case of the binary numeral system, like (A.3), define

$$d_i(x) := \lfloor 2^i x \rfloor - 2 \lfloor 2^{i-1} x \rfloor, \quad x \in [0, 1), \tag{A.4}$$

and we have

$$x = \sum_{i=1}^{\infty} 2^{-i} d_i(x), \quad x \in [0, 1).$$

Example A.4. How is 0.563 in the decimal numeral system described in the binary numeral system? Following (A.4), we see

$$\begin{aligned} d_1(0.563) &= \lfloor 2 \times 0.563 \rfloor - 2 \times \lfloor 0.563 \rfloor = 1 - 0 = 1, \\ d_2(0.563) &= \lfloor 4 \times 0.563 \rfloor - 2 \times \lfloor 2 \times 0.563 \rfloor = 2 - 2 = 0, \\ d_3(0.563) &= \lfloor 8 \times 0.563 \rfloor - 2 \times \lfloor 4 \times 0.563 \rfloor = 4 - 4 = 0, \\ d_4(0.563) &= \lfloor 16 \times 0.563 \rfloor - 2 \times \lfloor 8 \times 0.563 \rfloor = 9 - 8 = 1, \\ &\vdots \end{aligned}$$

and hence it is an infinite fraction $0.1001\ldots$. An equivalent easier method is

$$\begin{aligned} 2 \times 0.563 &= 1.126 = \underline{1} + 0.126, \\ 2 \times 0.126 &= 0.252 = \underline{0} + 0.252, \\ 2 \times 0.252 &= 0.504 = \underline{0} + 0.504, \\ 2 \times 0.504 &= 1.008 = \underline{1} + 0.008, \\ 2 \times 0.008 &= 0.016 = \underline{0} + 0.016, \end{aligned}$$

lining up underlined digits, we get the binary representation 0.10010....

Alternatively, describing 563/1000 in the binary numeral system as

$$\frac{1000110011}{1111101000},$$

we get the ratio using long division:

$$\begin{array}{r@{\,)\,}l}
 & 0.1001\ldots \\ \cline{2-2}
1111101000 & 1000110011.0 \\
 & 111110100\ 0 \\ \cline{2-2}
 & 111111\ 0000 \\
 & 111110\ 1000 \\ \cline{2-2}
\end{array}$$

Remark A.1. Recall that $1 = 0.9999\ldots$ in the decimal numeral system. Just like this, description of fractions in the binary numeral system is not necessarily unique. For example, $1/2 = 0.1 = 0.01111\ldots$. Adopting Definition (A.4) means that 1/2 should be described as 0.1 in the binary numeral system.

A.3 Limit of sequence and function

At high school, about the limit of sequences and functions, students learn, for example, $\lim_{n\to\infty} a_n = a$ as "If n gets larger and larger, a_n gets closer and closer to a". This description is too vague for advanced mathematics. Here we introduce the rigorous treatment of limit that was established by Cauchy, Weirestrass and others in the 19-th century.

A.3.1 *Convergence of sequence*

Let us consider quantitatively the situation "If n gets larger and larger, a_n gets closer and closer to a". Namely, we ask "How large shall we take n so that a_n can be how close to a?" The answer should be "If we take n greater than N, then the difference between a_n and a is less than $\varepsilon > 0$." Since the difference between a_n and a can be as small as we want, we may take any ε provided that it is positive. Therefore for any given $\varepsilon > 0$, if we can take an N that satisfies the condition "If n is greater than N, the difference between a_n and a is less than ε", we may well say that a_n converges to a as n tends to infinity.

Describing the above idea with only strictly selected words, we reach the following sophisticated definition.

Definition A.3. We say a real sequence $\{a_n\}_{n=1}^{\infty}$ is convergent to $a \in \mathbb{R}$ if for any $\varepsilon > 0$, there exists an $N \in \mathbb{N}_+$ such that $|a_n - a| < \varepsilon$ holds for any $n > N$. This is written as $\lim_{n\to\infty} a_n = a$.

In this definition, there is no phrase like "n gets greater and greater" or "a_n gets closer and closer to a", but it describes the situation admitting no possibility of misunderstanding.

Example A.5. Based on Definition A.3, let us prove

$$\lim_{n\to\infty} \frac{1}{\sqrt{n}} = 0. \tag{A.5}$$

First, take any $\varepsilon > 0$. We want to see

$$\left| \frac{1}{\sqrt{n}} - 0 \right| < \varepsilon \tag{A.6}$$

for large n. Solving this inequality in n, we get

$$n > \frac{1}{\varepsilon^2}.$$

Now, let $N := \lfloor 1/\varepsilon^2 \rfloor + 1$. Then, for any $n > N$, (A.6) holds. Thus we have proved (A.5).

Proposition A.1. *A convergent sequence* $\{a_n\}_{n=1}^{\infty}$ *is* bounded, *i.e., there exists an* $M > 0$ *such that* $|a_n| \leqq M$ *for any* $n \in \mathbb{N}_+$.

Proof. Let $\lim_{n\to\infty} a_n = a$. Then, (taking $\varepsilon = 1$) there exists an $N \in \mathbb{N}_+$ such that $|a_n - a| < 1$ for any $n > N$, in particular, $|a_n| < |a| + 1$ for any $n > N$. Now, let $M > 0$ be

$$M := \max\{\, |a_1|, |a_2|, \ldots, |a_N|, |a| + 1\,\}.$$

Then, for any $n \in \mathbb{N}_+$, we have $|a_n| \leqq M$. □

Proposition A.2. *Assume that* $\{a_n\}_{n=1}^{\infty}$ *converges to* 0. *Then, any* $A < B$, *it holds that*

$$\lim_{n\to\infty} \max_{n+A\sqrt{n} \leqq k \leqq n+B\sqrt{n}} |a_k| = 0. \tag{A.7}$$

Proof. By the assumption, for any $\varepsilon > 0$, there exists an $N \in \mathbb{N}_+$ such that for any $k > N$, we have $|a_k| < \varepsilon$. If $A \geqq 0$, then for $n > N$, we see

$$\max_{n+A\sqrt{n} \leqq k \leqq n+B\sqrt{n}} |a_k| \leqq \max_{n \leqq k \leqq n+B\sqrt{n}} |a_k| < \varepsilon,$$

which implies (A.7). If $A < 0$, when $n > N' := \lfloor 4A^2 \rfloor + 1$, we see

$$\frac{A}{\sqrt{n}} > \frac{A}{\sqrt{4A^2+1}} > \frac{A}{\sqrt{4A^2}} = \frac{A}{2|A|} = -\frac{1}{2},$$

and hence

$$n + A\sqrt{n} = n\left(1 + \frac{A}{\sqrt{n}}\right) > \frac{n}{2}.$$

Therefore for any $n > \max\{N', 2N\}$, we have $n + A\sqrt{n} > N$, which implies that

$$\max_{n+A\sqrt{n} \leqq k \leqq n+B\sqrt{n}} |a_k| \leqq \max_{N+1 \leqq k \leqq n+B\sqrt{n}} |a_k| < \varepsilon.$$

Thus (A.7) holds. □

Example A.6. Let us show that $\lim_{n\to\infty} a_n = a$ implies

$$\lim_{n\to\infty} \frac{a_1 + a_2 + \cdots + a_n}{n} = a. \tag{A.8}$$

First, take any $\varepsilon > 0$. By the assumption, there exists an $N_1 \in \mathbb{N}_+$ such that for any $n > N_1$, we have

$$|a_n - a| < \frac{\varepsilon}{2}. \tag{A.9}$$

On the other hand, since $\{a_n - a\}_{n=1}^{\infty}$ converges (to 0), it is bounded (Proposition A.1). Namely, there exists an $M > 0$ such that for any $n \in \mathbb{N}_+$, we have $|a_n - a| \leqq M$. Then, putting $N_2 := \lfloor 2N_1M/\varepsilon \rfloor + 1$, we see for any $n > N_2$,

$$\left|\frac{(a_1 - a) + \cdots + (a_{N_1} - a)}{n}\right| \leqq \frac{N_1M}{n} < \frac{N_1M}{N_2} < \frac{\varepsilon}{2}. \tag{A.10}$$

Finally, putting $N := \max\{N_1, N_2\}$, it follows from (A.9) and (A.10) that for any $n > N$, we have

$$\begin{aligned}
&\left|\frac{a_1 + a_2 + \cdots + a_n}{n} - a\right| \\
&\leqq \left|\frac{(a_1 - a) + \cdots + (a_{N_1} - a)}{n}\right| + \left|\frac{(a_{N_1+1} - a) + \cdots + (a_n - a)}{n}\right| \\
&< \frac{\varepsilon}{2} + \frac{|a_{N_1+1} - a| + \cdots + |a_n - a|}{n} \\
&< \frac{\varepsilon}{2} + \frac{\varepsilon}{2} \cdot \frac{n - N_1}{n} < \varepsilon.
\end{aligned}$$

This completes the proof of (A.8).

A.3.2 *Continuity of function of one variable*

Definition A.4. (i) We say a function f is convergent to r as x tends to a if for any $\varepsilon > 0$, there exists a $\delta > 0$ such that for any x satisfying $0 < |x - a| < \delta$, we have $|f(x) - r| < \varepsilon$. This is written as $\lim_{x \to a} f(x) = r$.
(ii) We say f is continuous at $x = a$ if $\lim_{x \to a} f(x) = f(a)$.
(iii) We say f is continuous in an interval $(a, b) := \{ x \mid a < x < b \}$ if for any $c \in (a, b)$, it is continuous at $x = c$.

These rigorous definitions of limit (Definition A.3 and Definition A.4) are called (ε, δ)-*definition.*

Example A.7. Let us show that $f(x) := x^2$ is continuous in the whole interval $\mathbb{R}$. First, take any $c \in \mathbb{R}$ and any $\varepsilon > 0$. We have to show the existence of a $\delta > 0$ such that

$$|f(x) - f(c)| = |x^2 - c^2| = |x - c||x + c|$$

becomes smaller than ε if $|x - c| < \delta$. Now, if $|x - c| < \delta$, we have

$$\begin{aligned} |x - c||x + c| &= |x - c||(x - c) + 2c| \\ &\leqq |x - c|(|x - c| + 2|c|) < \delta(\delta + 2|c|), \end{aligned}$$

and hence it is enough to find a $\delta > 0$ such that

$$\delta(\delta + 2|c|) < \varepsilon.$$

Let us solve this inequality in δ; adding $|c|^2$ to the both sides, we get

$$\delta^2 + 2|c|\delta + |c|^2 < |c|^2 + \varepsilon,$$

from which we derive

$$0 < \delta < \sqrt{|c|^2 + \varepsilon} - |c|. \tag{A.11}$$

Conversely, for any δ satisfying (A.11) and any x satisfying $|x - c| < \delta$, we see

$$|x^2 - c^2| < \varepsilon.$$

Thus $f(x) = x^2$ is continuous at $x = c$. Since c is any element of $\mathbb{R}$, it is continuous in $\mathbb{R}$.

A.3.3 *Continuity of function of several variables*

Definition A.4 naturally extends to the continuity of functions of several variables.

Definition A.5. We say a function of d variables $f : \mathbb{R}^d \to \mathbb{R}$ is continuous at a point $(a_1, \ldots, a_d) \in \mathbb{R}^d$ if for any $\varepsilon > 0$, there exists a $\delta > 0$ such that for any $(x_1, \ldots, x_d) \in \mathbb{R}^d$ satisfying $|x_1 - a_1| + \cdots + |x_d - a_d| < \delta$, we have

$$|f(x_1, \ldots, x_d) - f(a_1, \ldots, a_d)| < \varepsilon.$$

We say f is continuous in a domain $D \subset \mathbb{R}^d$ if it is continuous at each point of D.

For example, in the proof of Lemma 3.4, (3.53) is proved by the continuity of a function of 5 variables

$$f(x_1, x_2, x_3, x_4, x_5) := \frac{1}{\sqrt{4x_1x_2}} \cdot \frac{1 + x_3}{(1 + x_4)(1 + x_5)}$$

at a point $(\frac{1}{2}, \frac{1}{2}, 0, 0, 0)$. Similarly, for the proof of (3.49), we use, in the last paragraph of the proof of the lemma, the continuity of a function of two variables

$$g(x_1, x_2) := \exp(x_1)(1 + x_2) - 1$$

at the origin $(0, 0)$.

A.4 Limits of exponential function and logarithm

Proposition A.3.

(i) $\displaystyle\lim_{x \to \infty} x^a b^{-x} = 0, \quad a > 0,\ b > 1.$

(ii) $\displaystyle\lim_{x \to \infty} x^{-a} \log x = 0, \quad a > 0.$

(iii) $\displaystyle\lim_{x \to +0} x \log x = 0.$[†7]

Proof. (i) Let $c(x) := x^a b^{-x}$. We can take a large $x_0 > 0$ so that

$$0 < \frac{c(x_0 + 1)}{c(x_0)} = \left(1 + \frac{1}{x_0}\right)^a b^{-1} =: r < 1.$$

[7]This is called the *right-sided limit.* It means that the limit as x tends 0 *from above.* More precisely, it means that for any $\varepsilon > 0$, there exists a $\delta > 0$ such that for any x satisfying $0 < x < \delta$, we have $|x \log x| < \varepsilon$.

Then,

$$0 < \frac{c(x+1)}{c(x)} = \left(1+\frac{1}{x}\right)^{a} b^{-1} < r, \quad x > x_0,$$

and hence

$$0 < \frac{c(x+n)}{c(x)} = \frac{c(x+n)}{c(x+n-1)} \cdot \frac{c(x+n-1)}{c(x+n-2)} \cdots \frac{c(x+1)}{c(x)} < r^n, \quad x > x_0.$$

We therefore have

$$0 < c(x) < r^{\lfloor x-x_0 \rfloor} \times \max_{x_0 \leqq y \leqq x_0+1} c(y) \to 0, \quad x \to \infty.$$

(ii) By putting $y := \log x$, (i) implies that

$$\lim_{x\to\infty} x^{-a} \log x = \lim_{y\to\infty} e^{-ay} y = 0.$$

(iii) By putting $y := 1/x$, (ii) implies that

$$\lim_{x\to+0} x\log x = \lim_{y\to\infty} \frac{1}{y} \log\frac{1}{y} = -\lim_{y\to\infty} y^{-1}\log y = 0.$$

□

Proposition A.4. *For any $x > 0$,*

$$\lim_{n\to\infty} \frac{x^n}{n!} = 0.$$

Proof. Take $N \in \mathbb{N}_+$ so that $x < N/2$. Then, for $n > N$, we see

$$\begin{aligned} \frac{x^n}{n!} &= \frac{x^N}{N!} \times \frac{x}{N+1} \cdot \frac{x}{N+2} \times \cdots \times \frac{x}{n} \\ &< \frac{x^N}{N!} \times \left(\frac{x}{N}\right)^{n-N} \\ &< \frac{x^N}{N!} \times \left(\frac{1}{2}\right)^{n-N}, \end{aligned}$$

which converges to 0 as $n \to \infty$. □

A.5 C language program

Today, by the spread of computers, we can easily execute large-scale computations. It has led to an explosive expansion of concretely computable areas of mathematical theory. To make mathematical theory more useful in practice, readers should study computer basics.

The calculation of Example 4.4 was done by the following C-program ([Kernighan and Ritchie (1988)]). It outputs 546177 as the value of $S_{10^6}(g(\omega'))$ and the estimated value 0.546177 of the probability p.

```
/*==============================================================*/
/*  file name: example4_4.c                                     */
/*==============================================================*/
#include <stdio.h>

#define SAMPLE_NUM 1000000
#define M          100
#define M_PLUS_J   119

/* seed */
char xch[] =
   "1110110101" "1011101101" "0100000011" "0110101001"
   "0101000100" "0101111101" "1010000000" "1010100011"
   "0100011001" "1101111101" "1101010011" "111100100";
char ach[] =
   "1100000111" "0111000100" "0001101011" "1001000001"
   "0010001000" "1010101101" "1110101110" "0010010011"
   "1000000011" "0101000110" "0101110010" "010111111";

int  x[M_PLUS_J], a[M_PLUS_J];

void longadd(void) /* x = x + a (long digit addition) */
{
 int i, s, carry = 0;
 for ( i = M_PLUS_J-1; i >= 0; i-- ){
   s = x[i] + a[i] + carry;
   if ( s >= 2 ) {carry = 1; s = s - 2; } else carry = 0;
   x[i] =  s;
 }
}

int maxLength(void) /* count the longest run of 1's */
{
 int len = 0, count = 0, i;
 for ( i = 0; i <= M-1; i++ ){
  if ( x[i] == 0 )
  { if ( len < count ) len = count; count = 0;}
  else count++;  /*  if x[i]==1 */
 }
```

```
 if ( len < count ) len = count;
 return len;
}

int main()
{
 int n, s = 0;
 for( n = 0; n <= M_PLUS_J-1; n++ ){
  if( xch[n] == '1' ) x[n] = 1; else x[n] = 0;
  if( ach[n] == '1' ) a[n] = 1; else a[n] = 0;
 }
 for ( n = 1; n <= SAMPLE_NUM; n++ ){
  longadd();
  if ( maxLength() >= 6 ) s++;
 }
 printf( "s=%6d, p=%7.6f\n", s, (double)s/(double)SAMPLE_NUM);
 return 0;
}
/*================= End of example4_4.c =======================*/
```

List of mathematicians

Mathematician	Birth–Death	Related subject in this book
Euclid	B.C.3	—'s theorem
I. Newton	1642–1727	Motion equation
G.W. Leibniz	1646–1716	Binary numeral system
J. Bernoulli	1654–1705	—'s theorem
A. de Moivre	1667–1754	Limit of binary distribution
B. Taylor	1685–1731	—'s formula
C. Goldbach	1690–1764	—'s conjecture
J. Stirling	1692–1770	—'s formula
L. Euler	1707–1783	—'s integral
P.S. Laplace	1749–1827	Limit of binary distribution
A.-M. Legendre	1752–1833	— transformation
C.F. Gauss	1777–1855	—ian distribution
S.D. Poisson	1781–1840	Namer of law of large numbers
A.L. Cauchy	1789–1857	(ε, δ)-definition
K. Weierstrass	1815–1897	(ε, δ)-definition
P.L. Chebyshev	1821–1894	—'s inequality
A.A. Markov	1856–1922	—'s inequality
D. Hilbert	1862–1943	—'s 6-th problem
E. Borel	1871–1956	Normal number theorem
H. Lebesgue	1875–1941	Measure theory
H. Weyl	1885–1955	— transformation
G. Pólya	1887–1985	Namer of central limit theorem
H. Cramér	1893–1985	— -Chernoff's inequality
A.N. Kolmogorov	1903–1987	Axiom of probability theory
J. von Neumann	1903–1957	Monte Carlo method
K. Gödel	1906–1978	— number
S. Kleene	1909–1994	—'s normal form
S.M. Ulam	1909–1984	Monte Carlo method
A.M. Turing	1912–1954	— machine
H. Chernoff	1923–	Cramér- —'s inequality
R. Solomonoff	1926–2009	Random number
G. Chaitin	1947–	Random number

Further reading

This book exclusively dealt with limit theorems to emphasize the most important mission of probability theory—analysis of randomness. Of course, there are many other limit theorems not presented in this book, not necessarily determining events of probability close to 1. Anyhow, limit theorem is not the only theme of probability theory. To learn richness of probability theory, [Feller (1968)] and [Sinai (1992)] are recommended to read.

Before then, readers should master calculus and linear algebra. What follows are hints for those who have completed them.

In this book, we restricted the sample space Ω to be a finite set. To deal with infinite sample space, we need measure theory (§1.5.1). To study it, [Bilingsley (2012)] is recommended to read. Random number is merely an item of computation theory. Including it, to study computation theory, [Sipser (2012)] is recommended to read. About recursive function and algorithmic randomness, [Rogers (1967)], [Downey and Hirschfeld (2010)] and [Nies (2009)] are textbooks for graduate students and researchers. Books about the Monte Carlo method are really numerous. To study rigorous basic theory of it, [Sugita (2011)] is recommended to read. For advanced theory of it, [Bouleau and Lépingle (1994)] is recommended to read.

Bibliography

Billingsley, P. (2012). *Probability and Measure*, Anniversary edn. (Wiley).

Bouleau, N. and Lépingle, D. (1994). *Numerical methods for stochastic processes*, Wiley Series in Probability and Statistics (Wiley).

Downey, R. G. and Hirschfeld, D. R. (2010). *Algorithmic Randomness and Complexity* (Springer).

Feller, W. (1968). *An Introduction of Probability Theory and its Applications*, Vol. 1, 3rd edn. (Wiley).

Hardy, G. H. and Wright, E. M. (1979). *An introduction to the theory of numbers*, 5th edn. (Oxford Science Publications).

Kernighan, B. W. and Ritchie, D. M. (1988). *The C Programming Language*, 2nd edn. (Prentice Hall).

Kolmogorov, N. (1933). *Grundbegriffe der Wahrscheinlichkeitsrechnung, Ergebnisse der Mathematik und Ihrer Grenzgebiete* Vol. 2. English translation: *Foundation of the probability theory*, 2nd edn. (Chelsea (1956)).

Laplace, P. S. (1812). *Théorie analytique des probabilités.*

Li, M. and Vitányi, P. (2008). *An Introduction to Kolomogorov Complexity and Its Applications*, 3rd edn. (Springer).

Nies, A. (2009). *Computability and Randomness, Oxford Logic Guides* Vol. 51. (Oxford Univ. Press).

Rogers, H. Jr. (1967). *Theory of recursive functions and effective computability* (McGraw-Hill).

Sinai, Y. G. (1992). *Probability Theory—An Introductory Course* (Springer).

Sipser, M. (2012). *Introduction to the Theory of Computation*, 3rd edn. (Course Technology Inc.).

Sugita, H. (2011). *Monte Carlo Method, Random Number, and Pseudorandom Number, MSJ Memoirs* Vol. 25. (World Scientific).

Takahashi, M. (1991). *Theory of computation (Computability and λ-calculus)*, (Kindai-Kagaku sha (*Japanese*)).

Index

Printed in the United States
By Bookmasters